STP 1471

Titanium, Niobium, Zirconium, and Tantalum for Medical and Surgical Applications

Lyle D. Zardiackas, Matthew J. Kraay, and Howard L. Freese, editors

ASTM Stock Number: STP1471

ASTM
100 Barr Harbor Drive
PO Box C700
West Conshohocken, PA 19428-2959

Printed in the U.S.A.

Library of Congress Cataloging-in-Publication Data

ISBN: 0-8031-3497-5

Symposium on Titanium, Niobium, Zirconium, and Tantalum for Medical and Surgical
Applications (2004: Washington, DC)
Titanium, niobium, zirconium, and tantalum for medical and surgical applications / Lyle D.
Zardiackas, Matthew J. Kraay, and Howard L. Freese, editors.
    p. ; cm. — (STP ; 1471)
Includes bibliographical references and index.
ISBN 0-8031-3497-5
1. Metals in medicine—Congresses. 2. Metals in surgery—Congresses. 3. Alloys—Therapeutic
use—Congresses. 4. Titanium—Therapeutic use—Congresses. 5. Niobium—Therapeutic use—
Congresses. 6. Zirconium—Therapeutic use—Congresses. 7. Implants, Artificial—Congresses. 8.
Prostheses—Congresses.
    [DNLM: 1. Titanium—therapeutic use—Congresses. 2. Alloys—therapeutic use—Congresses. 3.
Niobium—therapeutic use—Congresses. 4. Prostheses and Implants—Congresses. 5. Tantalum—
therapeutic use—Congresses. 6. Zirconium—therapeutic use—Congresses. QT 37 T6176 2006]
I. Zardiackas, Lyle D. II. Kraay, Matthew J., 1955- III. Freese, Howard L., 1941- IV. ASTM
International. V. ASTM special technical publication; 1471.

R857.M37T58 2006
610.28—dc22

                                                                                    2005029979

Photocopy Rights

**Authorization to photocopy items for internal, personal, or educational classroom use, or
the internal, personal, or educational classroom use of specific clients, is granted by the
American Society for Testing and Materials International (ASTM) provided that the appropriate
fee is paid to the Copyright Clearance Center, 222 Rosewood Drive, Danvers, MA 01923; Tel:
978-750-8400; online: http://www.copyright.com/.**

Peer Review Policy

Each paper published in this volume was evaluated by two peer reviewers and at least one editor.
The authors addressed all of the reviewers' comments to the satisfaction of both the technical
editor(s) and the ASTM International Committee on Publications.
    The quality of the papers in this publication reflects not only the obvious efforts of the authors and
the technical editor(s), but also the work of the peer reviewers. In keeping with long-standing
publication practices, ASTM International maintains the anonymity of the peer reviewers. The ASTM
International Committee on Publications acknowledges with appreciation their dedication and
contribution of time and effort on behalf of ASTM International.

Printed in Ann Arbor, MI
Dec 2005

Foreword

This publication, *Titanium, Niobium, Zirconium, and Tantalum for Medical and Surgical Applications* includes peer reviewed papers presented at the ASTM F04 symposium by this same name in November of 2004. The symposium, held in Washington, DC, on November 9–10, 2005, focused on alloys whose primary constituents were one or more of these elements. The information included in the symposium was intended to provide an update on research results obtained since the last ASTM symposium on *Medical Applications of Titanium and Its Alloys* in 1994. The chairs of the symposium were Lyle D. Zardiackas from the University of Mississippi Medical Center, Howard Freese from Allvac, and Matthew Kraay from Case Western Reserve University and are likewise editors of this publication.

In light of the success of the previous symposium in 1994, the scope of this symposium was the presentation of information on the development of new alloys and processing techniques for medical applications, characterization of fundamental materials properties critical to their use for biomedical applications, and evaluation of biological and clinical performance.

The editors would like to express their appreciation to Dorothy Fitzpatrick for her tireless efforts in organizing the symposium, Maria Langiewicz and Don Marlowe for keeping us on track in publishing the papers and this text, and to Kathy Perrett for keeping the three of us organized and on time. Finally we would like to thank all of the ASTM staff for their efforts and the many reviewers of the individual papers for their time and expertise.

Lyle D. Zardiackas, Ph.D
Professor and Chair
Department of Biomedical Materials Science

Howard L. Freese, PE
Manager Business Development
Biomedical

Matthew J. Kraay, MS, MD
Associate Professor of Orthopaedics

Contents

v

BIOLOGICAL AND CLINICAL EVALUATION

Overview

The use of the reactive metals and their alloys for medical applications has continued to expand. Because of their unique properties, they have found use in a variety of biomedical applications from pace makers to hips. Since the time of the last ASTM symposium on medical applications of titanium, a great deal of research has been focused on the development, processing, properties and clinical performance of devices made from these alloys. As such, the symposium was divided into three sections 1) processing, 2) properties, and 3) biological and clinical performance.

Alloy Processing

The seven papers in this section include information on the formulation and processing of five important, new implantable metallic biomaterials:

- Ti-12Mo-6Zr-2Fe alloy ("TMZF")
- Zr-2.5Nb alloy ("Oxidized Zirconium", "Oxinium")
- Tantalum foam ("Trabecular Metal")
- Ti-35Nb-7Zr-5Ta alloy ("TiOsteum")
- Ti-15Mo alloy

The three titanium materials are metastable beta titanium alloys, manufactured and supplied in the mill annealed condition according to the guidelines of the ASTM F-04.12 "Metallurgical Materials" subcommittee. Five of these papers review the development and the processing of these beta titanium alloys, and the desired performance and properties of the alloy designers. Melting, thermomechanical processing, and finishing of semi-finished mill product forms can be quite different for these alloys than for the four α-phase CP titanium grades (CP-1, -2, -3, and -4), or for the three major $\alpha + \beta$ titanium alloys (Ti-6Al-4V, Ti-6Al-4V ELI, and Ti-6Al-7Nb). Although not discussed in detail in these papers, manufacturers of medical and surgical devices may experience initial difficulties as they establish cold forming and machining processes for these metastable β titanium alloys when they convert bar, rod, and wire products into finished device components. Several papers covered the influence of processing and chemical composition, particularly oxygen content, on the mechanical properties and microstructure of semi-finished mill products.

Papers by Murray et al. and Jablokov et al. look at chemical composition of ten different titanium grades and alloys. The Murray paper shows that Ti-12Mo-6Zr-2Fe alloy has a higher yield strength and better ductility, according to the published ASTM F 1813 standards, than the three major $\alpha + \beta$ titanium alloys (F 136, F 1295, and F 1472). Also, the F 1813 standard allows for a higher range of oxygen content, up to 0.28%, than those same three $\alpha + \beta$ alloys (0.13%, 0.20%, and 0.20% maximum respectively). The beta alloys generally have better ductility than $\alpha + \beta$ titanium alloys, while the increased interstitial oxygen content and greater alloy content (e.g., 12+6+2 vs. 6+4 and 6+7) generally result in greater yield strength. The Jablokov paper reports on the correlation of yield strength values that have been reported in mill certifications plotted versus oxygen content of many titanium grades and alloys. These surprisingly linear data fall into several clusters of titanium material types: the four CP titanium grades (α), three $\alpha + \beta$ titanium alloys, and three metastable beta titanium alloys. In the metallurgical literature, beta titanium alloys are characteristically reported to have a much broader range of interstitial oxygen solubility because of the physical metallurgy of the "bcc" crystalline structure. Qazi et al. report on aging studies for Ti-35Nb-7Zr-5Ta alloy with three oxygen contents, with some impressive fractography and electron microscopy analysis. Qazi concludes with the observation that increasing oxygen content in beta titanium alloys suppresses omega (ω) phase and promotes alpha (α) phase formation, finely dispersed, that results in a powerful alloy strengthening effect. Jablokov found that the ductility of Ti-35Nb-7Zr-5Ta is not negatively affected if strength is increased due to increasing oxygen content.

ASTM F 2066 specifies a single microstructure and condition for the Ti-15Mo beta titanium alloy: a wrought alloy with a "fully recrystallized beta phase structure" in the "beta annealed" condition. This unique binary alloy has moderate strength with high ductility when manufactured and supplied in this condition. Marquardt and Shetty, and Jablokov et al. suggest manufacturing and processing techniques that can significantly increase the strength and high cycle fatigue properties of the alloy, without a huge sacrifice in ductility. One processing method is an alpha/beta annealing process, and the other is a cold work reduction operation followed by an aging thermal treatment. The potential benefits from these processing techniques appear to warrant expansion of ASTM F 2066 standard beyond the current beta solution treated and quenched condition. In both studies, a broad range of microstructural conditions and associated mechanical properties are reported for the Ti-15Mo beta titanium alloy by altering rolling and/or drawing operations, and by selecting appropriate heat treatment procedures.

The Zr-2.5Nb alloy requires novel processing techniques by the alloy manufacturer and by the device manufacturer to achieve the uniform, extremely hard, and durable ceramic-like surface that is the key to this improved alloy for hard bearing designs. This is the first and, thus far, the only zirconium-base alloy that has been approved for clinical use for load-bearing articular components in orthopaedic applications. Hunter et al. describe the production processes for the manufacture of zirconium sponge (similar to that for making titanium sponge), alloy ingot, and semi-finished bar products. The special heat treatment process for converting ("transforming") a metal component's surface into a zirconium-oxide ceramic bearing is basically a gas-to-solid state diffusion process. This heat treatment creates a gradient between the zirconium alloy subsurface and the zirconium-oxide ceramic on the surface; a transition from metal to zirconium-oxide ceramic. The result is a stable, durable, low-friction bearing surface that has promising early clinical results for a low-wear couple with a UMHWPE counterface.

Medlin et al. report on the most interesting metallurgical advancement, this based on a porous "CP" tantalum foam product that can be attached to a metallic substrate, usually titanium, to fashion a "bimetallic" composite structure. Two methods are described to make a strong and durable bond between the two metallic components at the interface: a sintered powder process utilizing a sprayed titanium powder, and a diffusion bonding process. Data, sectioned photographs, and photomicrographs are presented that reveal a strong and quite durable bond between the two metal surfaces. Extensive mechanical and chemical testing has been carried out on a tantalum-titanium acetabular cup construct.

Session 2: Alloy Properties

Over the last three decades, the development of titanium and its alloys containing zirconium and niobium as well as the alloyed tantalum (small amounts of nitrogen, oxygen, and iron) for implant applications has continued to increase. Many of these alloys have unique properties such as a lower modulus of elasticity, better fatigue properties in a saline environment and enhanced biocompatibility. However, there continues to be gaps in available information and in our understanding not only of the properties but in the mechanisms of failure of many of these systems. The eight papers in this session covered bulk and surface properties of a number of different titanium alloys, which may be used for implants, as well as an evaluation of the fracture mechanisms.

The first two papers in this section were focused on dental applications of two new Japanese titanium alloys. The first paper by Okabe et al. compares the mechanical properties of castings of titanium with increased amounts of iron and nitrogen as compared to Grade 4 CP titanium. The evaluated mechanical properties, with the exception of elongation, were greater as compared to cast Grade 4 CP titanium. The second paper by Watanabe et al. focused on the effect of α-case on the flexural bond strength of dental composite to CP Ti and Ti-6Al-7Nb. As anticipated, the presence of α-case adversely affected the bond strengths of the composite to the α-case substrate.

The third and fourth papers reported results of the properties of two new titanium alloys. The corrosion, single cycle mechanical properties and fatigue, as well as biocompatibility of β-Ti-15Zr-4Nb-4Ta was the subject of this paper by Okazaki. Results of this study indicated that the alloy had promise for biomedical applications and that further research was justified. The paper by Niinomi et al. on β-Ti-29 Nb-13 Ta-4.6 Zr showed that under the conditions evaluated, this alloy has equal or greater tensile and fatigue properties compared to Ti-6Al-4V ELI. Additionally, the alloy showed super elastic behavior and good biocompatibility.

The last four papers in this session, determined and compared a variety of physical electrochemical and mechanical properties of four currently used titanium alloys with ASTM specifications. The first paper by Petersen et al. compared the substrate microstructure, surface oxide structure and thickness, and corrosion as a function of three surface treatments on CP titanium, Ti-6Al-4V, and Ti-15Mo-2.8Nb-0.2Si. Results showed no effect on substrate microstructure, no significant effect on oxide composition or thickness, and no variability in corrosion resistance as a function of surface treatment. The next two papers (Williamson et al. and Roach et al.) outlined the results of comparative studies on Grade 4 CP Ti, Ti-6Al-7Nb (α/β), Ti-6Al-4V ELI (α/β), and β-Ti-15Mo. The first paper compared the slow strain rate stress corrosion cracking (SCC) of these alloys with and without a notch. Results showed no differences in SCC in distilled water and Ringers solution in either smooth or notched samples and analysis of fracture surfaces showed no SCC morphology. The paper by Roach et al. compared tension-tension corrosion fatigue (CF) of these same alloys under the same conditions as the previous paper. The results of this research showed no effect of testing media on the fatigue of smooth or notched samples but a pronounced effect of the notch with a reduction in fatigue properties for all alloys evaluated. The final paper of the session (Zardiackas et al.) examined the effect of anodization on the CF and SCC of Grade 4 CP titanium with high oxygen content. Results showed no difference in any of the properties evaluated regardless of whether samples were anodized or not anodized.

Session 3: Biological and Clinical Evaluation

This session focused on the biological and clinical evaluation of titanium, niobium, zirconium, and tantalum used in the medical and surgical setting. Over the last two decades, contemporary hip replacements have universally incorporated at least some degree of modularity into their design. This most commonly involves the use of a Morse taper connection of the femoral head to the neck of the femoral stem; however, certain implants allow for independent fitting of the metaphyseal and diaphyseal areas of the femur with the use of connected modular segments. Although the advantages of modularity in hip replacement surgery are many, concern exists regarding the potential for fretting and crevice corrosion at these modular junctions. Paper #1 by Urban et al. reports the author's evaluation of 14 retrieved modular body (SROM) total hip stems removed at the time of revision surgery. Although corrosion of the modular junctions was frequently observed, particulate corrosion products were also found in the periprosthetic tissues. The results of this study demonstrate that in addition to the concerns about structural failure of the corroded modular tapers, particulate products of crevice corrosion in these devices can contribute to third-body articular surface wear and increased particulate burden in the periprosthetic tissues and resultant osteolysis.

Over the past 20 years, biologic fixation of joint replacement implants to the underlying bone has been shown to be a reliable alternative to cemented fixation with polymethylmethacrylate. Durable osseointegration can be achieved via bone ingrowth into porous surfaces consisting of sintered CoCr beads or diffusion bonded titanium fibermetal or via bone ongrowth on to plasma sprayed titanium or appropriate grit-blasted surfaces. Grit-blasted or corundumized surfaces have traditionally utilized bombardment of the substrate with aluminum oxide particles. This process results in these abrasive particles being embedded in the implant substrate with the potential for adverse effects on the implant interface (e.g., osteolysis) and increased third-body articular surface wear. Paper #3 entitled "Contamination-free, Grit blasted Titanium Surface" by Windler, Weber, and Rieder describes a new method of grit-blasting for implant surfaces, which uses iron particles, which unlike alumina, can be removed by chemical dissolution in nitric acid.

The problem of osteolysis has been a major focus of research in the area of joint replacement over the last 15 years. What we have commonly referred to as "cement disease" in the past, has now been clearly shown to in actuality be "particle disease." All particles, whether they are comprised of bone cement, corrosion products, or polyethylene, metal or even ceramic wear debris, can elicit an osteolytic response if present in sufficient numbers and of appropriate morphology (e.g., size, shape, and surface characteristic). The paper by Sprecher et al. describes a practical methodology to generate titanium and stainless steel wear debris particles for use in further study of the process of osteolysis and its treatment. The toxicity of soluble implant debris of zirconium and niobium containing alloys on human periprosthetic cell types was the subject of the research by Hallab et al. Implant alloys containing zirconium and niobium appear to induce a similar cellular response to other traditional alloys such as Ti-6Al-4V (ASTM F-138) and Co-Cr-Mo (ASTM F-75). This is reassuring considering the fact that these newer implant alloys are expected to be used with increased frequency in the future.

It is our hope that the information given in this text will serve as a resource for those working in this field and as reminder of the need to understand these materials, which will undoubtedly serve as the structural backbone for load bearing implants for decades to come. We must always be cognizant that we can never understand too much about the materials and devices that we place in our fellow man.

Lyle D. Zardiackas, Ph.D
Professor and Chair
Department of Biomedical Materials Science

Howard L. Freese, PE
Manager Business Development
Biomedical

Matthew J. Kraay, MS, MD
Associate Professor of Orthopaedics

ALLOY PROCESSING

Journal of ASTM International, September 2005, Vol. 2, No. 8
Paper ID JAI12774
Available online at www.astm.org

Naomi G.D. Murray,[1] *Victor R. Jablokov,*[2] *and Howard L. Freese*[3]

Mechanical and Physical Properties of Titanium-12Molybdenum-6Zirconium-2Iron Beta Titanium Alloy

ABSTRACT: Howmedica developed the Titanium-12Molybdenum-6Zirconium-2Iron beta titanium alloy (also known as TMZF® beta titanium alloy) in the late 1980s. Two patents were issued for this high strength, low modulus, ductile, biocompatible, titanium base alloy [1,2]. The mechanical properties for wrought machining and forging bar in ASTM F 1813 show higher tensile and yield strengths than Ti-6Al-4V ELI (ASTM F 136), the most widely used titanium grade for medical and surgical implant applications. According to the ASTM standard, the TMZF alloy also has improved ductility over Ti-6Al-4V ELI, which is to be expected when comparing beta and alpha+beta titanium alloys. TMZF alloy offers higher tensile strength and greater flexibility than Ti-6Al-4V alloy. Data are presented for semi-finished TMZF mill product forms in the solution-annealed condition. The influence of oxygen as an interstitial strengthening element is shown for Ti-12Mo-6Zr-2Fe and compared with common titanium alloys used in or being considered for use in the medical device industry.

KEYWORDS: metals (for surgical implants), orthopedic medical devices, titanium alloys, beta titanium alloys, alpha-beta titanium alloys, TiOsteum, osseointegration, mechanical strength, elastic modulus, titanium/titanium alloy, titanium/titanium alloys (for surgical implants), TMZF

Introduction

The rigorous chemical and mechanical requirements for materials used in implantable devices has spawned attempts to develop materials specific to these applications. To date, there are several materials which meet the needs for implantable devices [3]. These materials can be allocated into four distinct groups—stainless steels (iron-based), cobalt-based alloys, titanium-based alloys, and other specialty grade alloys. The properties of these alloys are described by several international standards such as ASTM [3], ISO [4], and UNS [5].

The identification of the need for materials specific to the medical device industry, i.e., biomaterials, is relatively recent compared to other industries such as the chemical, naval, and aerospace industries. In such industries, corrosion, heat, and fatigue resistance are critical. It was from these industries that the early pioneers in biomaterials borrowed materials technology and applied stainless steel and cobalt-based alloys to the implantable materials industry, testing each for tissue compatibility. Such materials are now considered reference materials for newly developed biomaterials [6]. Research into titanium and zirconium alloys quickly followed the initial use of these cross-industry materials.

Some of the largest advancements in materials technology have occurred in the development of titanium and titanium alloys. The wide variety of uses for titanium systems and the number of titanium alloys used in the implantable materials industry has grown rapidly in the past 15 years.

Manuscript received 20 December 2004; accepted for publication 1 March 2005; published September 2005.
Presented at ASTM Symposium on Titanium, Niobium, Zirconium, and Tantalum for Medical and Surgical Applications on 9-10 November 2004 in Washington, DC.
[1] Research Engineer, Stryker Orthopaedics, Mahwah, NJ 07430.
[2] Senior Engineer, ATI Allvac, an Allegheny Technologies company, Monroe, NC 28110.
[3] Business Development Manager, ATI Allvac, an Allegheny Technologies company, Monroe, NC 28110.

The "Metallurgical Materials" Subcommittee, ASTM F4.12, has developed six new ASTM Standards since the early 1990s, and all have been balloted and approved by the "Medical and Surgical Materials and Devices" Main Committee, F04. These titanium alloys are listed in Table 1.

TABLE 1—*Listing of titanium alloys and their associated ASTM standards.*

Common Name	ASTM & ISO Standards	Microstructure	UNS Designation
Ti-6Al-7Nb Alloy ("TAN")	ASTM F 1295, ISO 5832-11	A+β	R56700
Ti-6Al-4V Alloy	ASTM F 1472, ISO 5832-3	A+β	R56400
Ti-13Nb-13Zr Alloy	ASTM F 1713	Metastable β	R58130
Ti-12Mo-6Zr-2Fe Alloy ("TMZF")	ASTM F 1813	Metastable β	R58120
Ti-15Mo Alloy	ASTM F 2066	Metastable β	R58150
Ti-3Al-2.5V Alloy (tubing only)	ASTM F 2146	A+β	R56320
Ti-35Nb-7Zr-5Ta Alloy ("TiOsteum")	Sub. F-04.12.23	Metastable β	R58350

Due to the outstanding corrosion resistance and good mechanical properties, Ti-6Al-4V (ASTM F1472, UNS R56400) is still one of the most widely used alloys in the orthopedic industry.   Downsides to this "work horse" alloy include notch sensitivity under fatigue conditions and relatively poor wear resistance.   Furthermore, the presence of aluminum and vanadium in the alloy and, therefore, in the debris generated under conditions of extreme wear have led to potential safety concerns since the early 1980s [7,8].

In response to such concerns, aluminum- and vanadium-free titanium alloys have been developed for use in the implantable materials industry. Vanadium-free alloys include the development of alloys such as Ti-5Al-2.5Fe and Ti-6Al-7Nb (ASTM F 1295, UNS R56700), the first new α+β metallic biomaterials. These alloys have mechanical properties similar to Ti-6Al-4V, but they have the benefit of the use of iron or niobium instead of vanadium as a beta stabilizing alloying element [7,8].

Following the development of the α+β alloys was the development of new metastable β-titanium alloys. The standards for Ti-13Nb-13Zr (ASTM F 1713) [9,10] and Ti-12Mo-6Zr-2Fe (ASTM F 1813) [1-3,11] were designed specifically for use in structural orthopedic implant applications. These alloys traveled through the ASTM subcommittee simultaneously. ASTM F 2066 was subsequently developed for Ti-15Mo [12]. Finally, ASTM F 2146 was developed to cover the α+β Ti-3Al-2.5V alloy tubing due to its good hot and cold workability as compared to the other α+β titanium alloys (Ti-6Al-4V ELI, Ti-6Al-7Nb, and Ti-6Al-4V).

The main focus of the following paper will be on the development of the β-titanium alloy, Ti-12Mo-6Zr-2Fe, focusing on the physical and mechanical properties of the alloy, considering its uses for structural orthopedic applications.

Development of Ti-12Mo-6Zr-2Fe

Due to the perceived safety concerns from the presence of vanadium in the most widely used titanium biomaterial, Ti-6Al-4V ELI, an alternative titanium alloy for use as a biomaterial was sought. Additionally, finite element studies suggested that the use of a low modulus material in prosthetic hips might better mimic the natural bone, allowing for more efficient distribution of stresses to the adjacent host bone during use [13,14]. These results, coupled with early results from animal studies suggesting that bone resorption is lessened by the presence of a lower modulus prosthesis [15,16], generated much interest in the development of low-modulus alloys to for structural orthopedic applications [17].

Although pure titanium has a hexagonal close packed (HCP) lattice (α), β-titanium alloys have a body centered cubic (BCC) titanium lattice structure and are more readily formable than their α+β-titanium alloy counter-parts. Additionally, β-titanium alloys generally have good tensile ductility, and the fracture toughness can be higher and the notch sensitivity lower than for an α+β alloy with a similar yield strength [18]. The development of Ti-12Mo-6Zr-2Fe by Wang, Gustavson, and Dumbleton yielded a low-modulus, high-strength, corrosion-resistant, biocompatible alloy with good wear and notch fatigue resistance [17].

Alloy Microstructure

The microstructure of Ti-12Mo-6Zr-2Fe is shown in Fig. 1. The β structure is retained entirely when cooled rapidly from temperatures above the β-transus (>730°C) and looks remarkably like the microstructure of pure titanium. However, Ti-12Mo-6Zr-2Fe is a metastable β-titanium alloy, and upon aging fine α-phase precipitates will form. The precipitation of the α-phase increases the modulus of the alloy, and therefore the single-phase β-alloy is preferred for structural orthopedic implants.

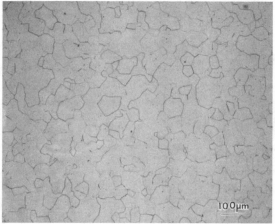

FIG. 1—*Representative micrograph showing the microstructure of Ti-12Mo-6Zr-2 Fe.*

Chemistry of Ti-12Mo-6Zr-2Fe

The nominal specification chemistry limits for various titanium alloys are listed in Table 2, showing the minimum and maximum allowable limits for metallic elements and interstitial gas elements. These data demonstrate numerically the differences between the four chemically pure (CP) titanium grades (alpha phase microstructure), the three α+β titanium alloys, and three metastable β titanium alloys.

TABLE 2—*Specification chemistry limits for commercially pure, alpha + beta and metastable beta titanium alloys* [11,12,19–22].

Element	Ti-CP-4 [a] ASTM F 67 Gr. 4 (min)	(max)	Ti-6Al-4V ELI ASTM F 136 (min)	(max)	Ti-6Al-7Nb ASTM F 1295 (min)	(max)	Ti-6Al-4V ASTM F 1472 (min)	(max)	Ti-12Mo-6Zr-2Fe ASTM F 1813 (min)	(max)	Ti-15Mo ASTM F 2066 (min)	(max)	Ti-35Nb-7Zr-5Ta F 04.12.23 (min)	(max)
Nitrogen		0.05		0.05		0.05		0.05		0.05		0.05		0.02
Carbon		0.08		0.08		0.08		0.08		0.05		0.10		0.02
Hydrogen		0.015		0.012		0.009		0.015		0.020		0.015		0.020
Oxygen		0.40		0.13		0.20		0.20		0.28		0.20		0.75
Iron		0.50		0.25		0.25		0.30	1.50	2.50		0.10		0.25
Aluminum			5.50	6.50	5.50	6.50	5.50	6.75						
Vanadium			3.50	4.50			3.50	4.50						
Niobium					6.50	7.50							34.00	37.00
Molybdenum									10.00	13.00	14.00	16.00		
Zirconium									5.00	7.00			6.30	8.30
Tantalum						0.50							4.50	6.50
Yttrium								0.005						
[b] Titanium (by Δ)	100.00	98.96	91.00	88.478	88.00	84.91	91.00	88.10	83.49	77.10	86.00	83.54	55.20	47.14
Total	100.00	100.00	100.00	100.00	100.00	100.00	100.00	100.00	100.00	100.00	100.00	100.00	100.00	100.00
[c] Titanium, ave.	99.48%		89.74%		86.46%		89.55%		80.30%		84.77%		51.17%	
Ave. Alloy Content	0.52%		10.26%		13.54%		10.45%		19.70%		15.23%		48.83%	

[a] ASTM F 67 provides for four different compositions: Grade 1 ("CP-1"), Grade 2 ("CP-2"), Grade 3 ("CP-3"), and Grade 4 ("CP-4").
[b] Titanium content is determined by difference for values in both the minimum ("min") and the maximum ("max") columns.
[c] The "Titanium, ave." value is the arithmetic average of the theoretical minimum and maximum limits of titanium content according to the appropriate ASTM standard.

An artificial but meaningful measure of the alloy content is obtained by calculating the "Titanium, average" value, which we define as the arithmetic average of the theoretical minimum and maximum limits of titanium content (by difference), according to the appropriate ASTM standard. Subtracting this value from unity, we obtain a measure of the alloy content (which includes interstitials). The metastable β-titanium alloy Ti-12Mo-6Zr-2Fe is one of the more highly alloyed titanium alloys within the list, only behind Ti-35Nb-7Zr-5Ta. Interestingly, these two alloys also have the largest allowable range for the amount of interstitial oxygen. The amount of oxygen in Ti-12Mo-6Zr-2Fe can range from 0.008–0.28 wt% and remain within the specification limits, while the specification limit for Ti-35Nb-7Zr-5Ta allows for up to 0.75 wt% oxygen.

Alloy Physical and Mechanical Properties

Physical Properties

The physical properties of Ti-12Mo-6Zr-2Fe and other titanium alloys are listed in Table 3. The average thermal expansion coefficient was found to be 8.8 × 10^{-6}/°C between 25°C and 250°C and 11.5 × 10^{-6}/°C between 525°C and 900°C. Between 250°C and 525°C, a departure from linearity is observed, which is indicative of the α/β phase transformation where upon heating, the α-phase transforms into the β-phase [17]. While the density of Ti-12Mo-6Zr-2Fe (5.0 g/cm^3) is higher than that of CP-Ti (4.51 g/cm^3) and Ti-6Al-4V (4.43 g/cm^3), it is similar to that of Ti-15Mo (4.95 g/cm^3) and much lower than other structural alloys used in the orthopedic industry, such as stainless steels and cobalt-based alloys. The β-transus of pure titanium is nominally 882°C, but it is sensitive to both metallic and interstitial alloying element. Due to the presence of the β-stabilizers Mo, Zr, and Fe, the β-transus of the metastable β-titanium alloys are generally lower than that of CP-Ti or pure titanium. The β-transus of Ti-12Mo-6Zr-2Fe is reported to be between 732°C and 754°C. The β-transus of Ti-6Al-4V is much higher than that of pure titanium (980°C versus 882°C, respectively).

TABLE 3—*Typical physical properties of common titanium alloys used for structural orthopedic applications* [17,18,23].

Property	CP-Ti Grade 4 ASTM F67	Ti-6Al-4V ELI ASTM F136	Ti-12Mo-6Zr-2Fe* ASTM F1813	Ti-15Mo ASTM F2066
Thermal Expansion Coefficient (x10^{-6}/°C)	8.9 (20–100°C) 10.1 (1000°C)	8.6 (20–100°C) 9.7 (20–650°C)	8.8 (25–250°C) 11.5 (525–900°C)	7.7 (20–275°C) 11.4 (455–550°C)
Beta Transus (°C)	950	980	732-754	~774
Density (g/cm^3)	4.5	4.43	5.0	4.95

*solution-treated condition.

General Mechanical Properties

Typical room temperature mechanical properties for common titanium alloys used in structural orthopedic applications are shown in Table 4. This table shows that Ti-12Mo-6Zr-2Fe has a higher yield strength and ultimate tensile strength than CP-Ti, Ti-6Al-4V, and Ti-15Mo. Despite having higher tensile strengths than Ti-6Al-4V, Ti-12Mo-6Zr-2Fe has high tensile ductility, as is observed by comparing the EL% and RA% for the alloys in Table 4. Additionally, the elastic modulus of Ti-12Mo-6Zr-2Fe is significantly lower than the traditional Ti-6Al-4V alloy (~80 GPa for Ti-12Mo-6Zr-2Fe as compared ~110 for Ti-6Al-4V). The smooth and notched (using a stress concentration factor of 1.6) rotating fatigue properties of Ti-12Mo-6Zr-2Fe as compared against Ti-6Al-4V are shown in Fig. 2. Ti-12Mo-6Zr-2Fe shows similar smooth fatigue strength to Ti-6Al-4V and, when using a run-out of 10×10^6 cycles, a fatigue stress of 585 MPa has been found [17]. However, Ti-12Mo-6Zr-2Fe has been found to be much less notch-sensitive than Ti-6Al-4V, having a notched strength of ~410 MPa (10^7 cycles) as compared to 210 MPa (10^7 cycles) for Ti-6Al-4V. This corresponds to a drop of only 47 % for Ti-12Mo-6Zr-2Fe compared to a drop in fatigue strength of 70 % for Ti-6Al-4V from the smooth fatigue strength [17]. The advantage of an improved notch fatigue resistance translates into a material that is more tolerant to surface stress concentrations during service.

TABLE 4—*Typical mechanical properties of common titanium alloys used for sructural orthopedic applications* [11,12,17–20,24,25].

Property	CP-Ti Grade 4 ASTM F 67	Ti-6Al-4V ELI ASTM F 136	Ti-12Mo-6Zr-2Fe* ASTM F 1813	Ti-15Mo** ASTM F 2066
YS (MPa)	480	850–900	1000–1060	483
UTS (MPa)	550	960–970	1060–1100	690
EL (%)	>15	10–15	18–22	20
RA (%)	30	35–47	64–73	60
E (GPa)	110	110	74–85	78

*Solution-annealed condition.
**β-annealed condition.

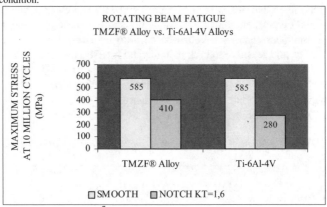

FIG. 2—*Maximum stress at 10^7 cycles for smooth and notched fatigue testing of Ti-12Mo-6Zr-2Fe and Ti-6Al-4V.*

The fracture toughness, a measure of the ability of a material to resist the propagation of a crack under load, of β-annealed Ti-12Mo-6Zr-2Fe was found to be higher than that of mill-annealed Ti-6Al-4V ELI (90 MPa√m as compared to 52 MPa√m, respectively) when tested in accordance with ASTM E 399 (Standard Test Method for Plane-Strain Fracture Toughness of Metallic Materials) [17]. Generally, materials with higher fracture toughness are less sensitive to damage in service.

Relationship Between Chemistry and Mechanical Properties for Titanium-Based Alloys
It is well known that the four grades of CP-Ti have slightly varying chemistries, and as the alloying content increases, particularly the amount of oxygen in the titanium, the strength

increases while ductility decreases [18]. Specifically, increasing the oxygen content from 0.18 % in CP-Ti Grade 1 to 0.40 % in CP-Ti Grade 4 increases the yield strength from 172 MPa to 482 MPa, respectively. Unfortunately, this increase in strength is accompanied by a decrease in ductility, where Grade 1 CP-Ti has an elongation of 24 %, while CP-Ti Grade 4 has an elongation of 15 %. A similar phenomenon is observed when comparing Ti-6Al-4V (maximum oxygen content of 0.20 %) with Ti-6Al-4V ELI (maximum oxygen content of 0.13 %), where the former has a yield strength of 860 MPa and the latter a yield strength of 795 MPa.

The commercially pure grades of titanium and the α+β titanium alloys are generally constrained to low oxygen contents. However, the metastable β-titanium alloys typically have higher alloying content, and some allow for higher oxygen content. This is true for alloys designed for use in the medical device industry. For example, Ti-12Mo-6Zr-2Fe (UNS R58120) allows for an alloying elements content of 20 % with a maximum oxygen content of 0.28 %, and Ti-35Nb-7Zr-5Ta (R58350) has 47 % alloying elements and 0.75 % proposed maximum oxygen content. Note that the maximum oxygen content for Ti-15Mo (UNS R58150) has been set at 0.20 %. It is thought that this may have been set a level lower than required; such a limit on the maximum oxygen content is more consistent with α+β titanium alloys.

Correlation of Oxygen Content with Alloy Mechanical Properties

Due to the higher allowable oxygen levels in some of the metastable β-titanium alloys, it is useful to understand the relationship between the oxygen content and the mechanical properties of the alloy. It has been possible for the authors to investigate such a correlation using analytical data from production laboratory systems for CP-Ti, α+β, and metastable β-titanium alloys used in both the aerospace and medical fields. A rigorous search of proprietary laboratory files for seven titanium alloys in the same condition and processed using similar or identical equipment was performed. Compositions investigated, including allowable oxygen content range, are listed in Table 5, and the relationship between oxygen content on the yield strength for these titanium alloys in the mill-annealed condition is shown in Fig. 3a. The same data for Ti-12Mo-6Zr-2Fe and Ti-35Nb-7Zr-5Ta are shown in Fig. 3b in order to observe more closely the relationship between the mill-annealed yield strength and ingot oxygen content for metastable β-titanium alloys. Data are shown in Table 6. All tested conditions were from round bar product, but had varying diameters. All were either Plasma Arc Melted or Vacuum Arc Melted, pressed, and rotary forged to intermediate billet, hot rolled to round bar/coil and finish-machined. All specimens came from material that conforms to the applicable biomedical specification. Each data point represents an average yield strength at a specific oxygen content, but each point may have come from one or more ingots/heats. Over 2000 data points have been analyzed to generate Fig. 3.

TABLE 5—*Specification average titanium and oxygen contents for eight titanium grades.*

Common Name	UNS Designation	Oxygen, min.	Oxygen, max.	Oxygen, ave.	Titanium, ave.
Ti CP1-4	R50700	0.0	0.40	0.20	99.48
Ti-6Al-4V ELI	R56401	0.0	0.13	0.065	89.74
Ti-6Al-7Nb	R56700	0.0	0.20	0.10	86.46
Ti-6Al-4V	R56400	0.0	0.20	0.10	89.55
Ti-12Mo-6Zr-2Fe	R58120	0.008	0.28	0.144	80.30
Ti-15Mo	R58150	0.0	0.20	0.10	84.77
Ti-35Nb-7Zr-5Ta	R58350	0.0	0.75	0.375	51.17

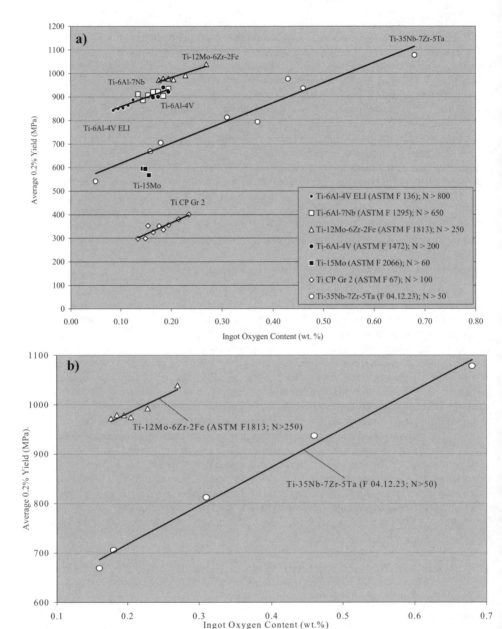

FIG. 3—*Average yield strength as function of ingot oxygen content showing (a) global relationship for CP-Ti, α+β titanium alloys and metastable β-titanium alloys and (b) relationship between two metastable β-titanium alloys, Ti-12Mo-6Zr-2Fe and Ti-35Nb-7Zr-5Ta.*

TABLE 6—*Average yield strength data as a function of ingot oxygen content for commercially pure, alpha + beta and metastable beta titanium alloys.*

Ti-35Nb-7Zr-5Ta		Ti-15Mo		Ti-12Mo-6Zr-2Fe		Ti CP Gr 2	
Ingot O (wt. %)	Ave. Yield (MPa)	Ingot O (wt. %)	Ave. Yield (MPa)	Ingot O (wt. %)	Ave. Yield (MPa)	Ingot O (wt. %)	Ave. Yield (MPa))
0.16	669	0.14	596	0.18	972	0.14	297
0.18	706	0.15	594	0.19	979	0.15	299
0.31	813	0.16	568	0.20	978	0.16	353
0.46	937			0.21	974	0.17	325
0.68	1078			0.23	992	0.18	352
				0.27	1038	0.19	336
						0.20	356
						0.22	381
						0.24	401

Ti-6Al-7Nb		Ti-6Al-4V		Ti-6Al-4V ELI	
Ingot O (wt. %)	Ave. 0.2 % Yield (MPa)	Ingot O (wt. %)	Ave. 0.2 % Yield (MPa)	Ingot O (wt. %)	Ave. 0.2 % Yield (MPa)
0.14	911	0.17	897	0.09	843
0.15	886	0.18	901	0.10	850
0.16	907	0.19	940	0.11	853
0.17	921	0.20	921	0.12	864
0.18	922	0.21	911	0.13	887
0.19	904	0.22	951	0.15	963
0.20	934				

The trends in Fig. 3 illustrate the well-documented effect of oxygen acting as an interstitial strengthener for titanium and its alloys. This results in a trend of increasing yield strength with increasing ingot oxygen content [26,27] for all alloys (Fig. 3) and spans a wide range of ingot oxygen levels. The exception is Ti-15Mo, which was tested only within a small range of oxygen contents due to the constraints in allowable oxygen content. For all alloys tested, the yield strength increases with increasing oxygen content. However, the oxygen content range for most alloys is small compared to that shown in Fig. 3 for Ti-35Nb-7Zr-5Ta. Therefore, it cannot be determined from the current data if the observed trends for CP-TiGr2(α) and the $\alpha + \beta$ values would continue.

Within the oxygen levels shown for Ti-12Mo-6Zr-2Fe (0.18–0.27 wt%), this alloy has the highest yield strength among the titanium alloy shown in Fig. 3. All other titanium alloys, both $\alpha+\beta$ and metastable β-titanium alloys shown, have yield strengths lower than Ti-12Mo-6Zr-2Fe, including metastable β-titanium alloy Ti-35Nb-7Zr-5Ta. However, when these two metastable β-titanium alloys are compared (Fig. 3b), it is observed that the relationship between yield strength and ingot oxygen content for both alloys is similar (similar slope). That is, although a wider range of ingot oxygen content has been studied for Ti-35Nb-7Zr-5Ta than for Ti-12Mo-6Zr-2Fe (0.16–0.68 compared to 0.18–0.27, respectively), the two curves appear to be roughly parallel. Furthermore, assuming that the relationship shown for Ti-12Mo-6Zr-2Fe alloy in Fig. 3b continues in the same manner with increasing oxygen content, the increase of approximately 60 % in yield strength observed for Ti-35Nb-7Zr-5Ta may be expected from Ti-12Mo-6Zr-2Fe. Further work is necessary to verify if such a relationship is valid at higher ingot oxygen content.

It is important to realize, however, that the relationship between yield strength and ingot oxygen content does not provide a complete picture of the relationship between alloy mechanical properties and oxygen concentration. In an effort to better understand this relationship, the effect of oxygen content (for the range of ingot oxygen content shown in Fig. 3) on the ultimate tensile strength, yield strength, percent elongation, and percent reduction of area during room temperature tensile tests is shown in Fig. 4*a* and *b* for Ti-12Mo-6Zr-2Fe and Ti-35Nb-7Zr-5Ta, respectively. As for Fig. 3, data shown are an average of available mill-annealed data for a specific oxygen content.

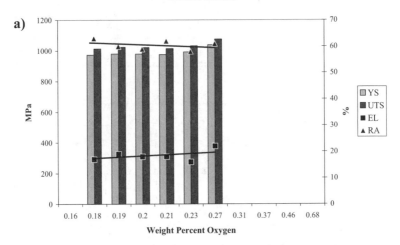

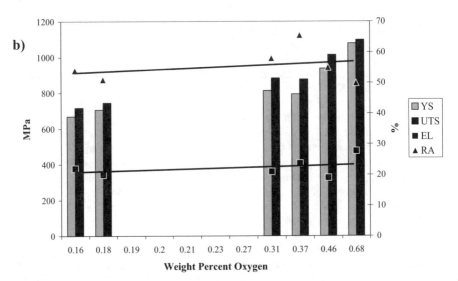

FIG. 4—*Yield strength (YS), ultimate tensile strength (UTS), elongation (EL), and area reduction (RA) versus ingot oxygen content for (a) Ti-12Mo-6Zr-2Fe and (b) Ti-35Nb-7Zr-5Ta in the mill-annealed condition.*

Similar to Fig. 3, Fig. 4*a* shows that, as expected, the ultimate tensile strength for Ti-12Mo-6Zr-2Fe also increases with increasing oxygen level. Ti-12Mo-6Zr-2Fe shows an increase in ultimate tensile strength from 1013 MPa to 1075 Mpa, with a change in oxygen content from 018 to 0.27 wt%. A similar trend is shown when comparing the data from Ti-12Mo-6Zr-2Fe to a wider range of oxygen content studied for Ti-35Nb-7Zr-5Ta (Fig. 4*b*). Of particular interest for both alloys is that the ductility does not show a significant decrease with increasing alloy strength and ingot oxygen content. This confirms observations reported in other publications [28].

Conclusions

The impetus, development, and properties of the metastable β-titanium alloy, Ti-12Mo-6Zr-2Fe, were reviewed, and the effect of oxygen concentration on alloy mechanical properties were discussed. The following conclusions can be made:

- The development of the biomaterial Ti-12Mo-6Zr-2Fe addressed many critical issues associated with implantable materials.
- The composition of the metastable β-titanium alloy, Ti-12Mo-6Zr-2Fe, avoids controversial alloying elements such as aluminum and vanadium.
- The alloy is seen to have an increased strength combined with a lower elastic modulus and increased fracture toughness as compared to the workhorse biomaterial Ti-6Al-4V ELI.
- Yield strength and ultimate tensile strength of Ti-12Mo-6Zr-2Fe and other titanium alloys increases with increasing ingot oxygen content.
- Within the range studied, the ductility of Ti-12Mo-6Zr-2Fe is not significantly affected by increasing oxygen content. However, this effect should be studied over a larger range of ingot oxygen levels.

References

[1]   Wang, K.G., J. and Dumbleton, J., *High Strength, Low Modulus, Ductile Biocompatible Titanium Alloy*, Pfizer Hospital Products Group, Inc., USA, Patent #4,857,269, 1989.

[2]   Wang, K., Gustavson, J., and Dumbleton, J., *Method of Making High Strength, Low Modulus, Ductile Biocompatible Titanium Alloy*, Pfizer Hospital Products Group, Inc., USA, Patent #4,952,236, 1990.

[3]   Lemons, J. and Freese, H., *Metallic Biomaterials For Surgical Implant Devices*, BoneZone, 2002, pp. 5–9.

[4]   *International Organization for Standardization*, ISO Central Secretariat, 1 rue de Varembe, Case postale 56, CH-1211.

[5]   *Metals and Alloys in the Unified Numbering System (UNS), Ninth Edition*, ASTM International, West Conshohocken, PA.

[6]   ASTM Standard F 763, "Standard Practice for Short-Term Screening of Implant Materials," ASTM International, West Conshohocken, PA, originally published as F 763-82.

[7]   Zwicker, R., et al., "Mechanical Properties and Tissue Reactions of a Titanium Alloy for Implant Material," *Titanium '80. Science and Technology Proceedings of the 4th*

International Conference on Titanium, The Metallurgical Society AIME, Kyoto, Japan, 1980.

[8]   Semlitsch, F., Staub, F., and Webber, H., "Titanium-Aluminum-Niobium Alloy, Development for Biocompatible, High Strength Surgical Implants," *Biomedical Technik*, Vol. 30, 1985, pp. 334–339.

[9]   Ahmed, T. and Rack, H. J., *Low Modulus Biocompatible Titanium Base Alloys for Medical Devices*, 1999.

[10] ASTM Standard F 1713, "Standard Specification for Wrought Titanium-13Niobium-13Zirconium Alloy for Surgical Implant Applications," ASTM International, West Conshohocken, PA.

[11] ASTM Standard F 1813, "Standard Specification for Titanium-12Molybdenum-6Zirconium-2Iron Alloy for Surgical Implant Applications," ASTM International, West Conshohocken, PA.

[12] ASTM Standard F 2066, "Standard Specification for Wrought Titanium-15Molybdenum for Surgical Implant Applications," ASTM International, West Conshohocken, PA.

[13] Cheal, E., Spector, M., and Hayes, W., "Role of Loads and Prosthesis Material Properties on the Mechanics of the Proximal Femur After Total Hip Arthroplasty," *Journal of Orthopedic Research*, Vol. 10, 1992, pp. 405–422.

[14] Prendergast, P. and Taylor, D., "Stress Analysis of the Proximo-Medial Femur After Total Hip Replacement," *Journal of Biomedical Engineering*, Vol. 12, No. 5, 1990, pp. 379–382.

[15] Bobyn, J. D., et al., "The Effect of Stem Stiffness on Femoral Bone Resorption After Canine Porous-Coated Total Hip Arthroplasty," *Clinical Orthopedics and Related Research*, 261, 1990, pp. 196–213.

[16] Bobyn, J. D., et al., "Producing and Avoiding Stress Shielding: Laboratory and Clinical Observation of Noncemented Total Hip Arthroplasty," *Clinical Orthopedics and Related Research*, 1992, 274, pp. 79–96.

[17] Wang, K., Gustavson, L., and Dumbleton, J., "Low Modulus, High Strength, Biocompatible Titanium Alloy for Medical Implants," *Titanium '92*, TMS, 1992.

[18] Donachie, M., *Titanium: A Technical Guide*, ASM International, 381, Materials Park, OH, 2000.

[19] ASTM Standard F 67, "Standard Specification for Unalloyed Titanium for Surgical Implant Devices," ASTM International, West Conshohocken, PA.

[20] ASTM Standard F 136, "Standard Specification for Wrought Titanium-6Aluminum-4Vanadium ELI (Extra Low Interstitial) Alloy for Surgical Implant Applications," ASTM International, West Conshohocken, PA.

[21] ASTM Standard F 1295, "Standard Specification for Wrought Titanium-6Aluminum-7Niobium Alloy for Surgical Implant Applications," ASTM International, West Conshohocken, PA.

[22] ASTM Standard F 1472, "Standard Specification for Wrought Titanium-6Aluminum-4Vanadium Alloy for Surgical Implant Applications," ASTM International, West Conshohocken, PA.

[23] *MatWeb, Material Property Data*, Automation Creation, Inc., http://www.matweb.com/.

[24] Zardiackas, L., Mitchell, D. W., and Disegi, J. A., "Characterization of Ti-15Mo Beta Titanium Alloy for Orthopedic Implant Applications," *Medical Applications of Titanium and its Alloys: The Material and Biological Issues, ASTM STP 1272*, S. A. Brown and J. E. Lemons, Eds., ASTM International, West Conshohocken, PA, 1996, pp. 60–75.

[25] Disegi, J. A., "Titanium Alloys for Fracture Fixation Implants," *Injury, International Journal of the Care of the Injured*, Vol. 31, 2000, p. S-D14-17.

[26] Imam, M. A. and Feng, C. R., "Role of Oxygen on Transformation Kinetics of Timetal-21S Titanium Alloy," *Advances in the Science and Technology of Titaniumn Alloy Processing*, TMS, Warrendale, PA, 1996.

[27] Qazi, J. I., et al., "The Effect of Duplex Aging on the Tensile Behavior of Ti-35Nb-7Zr-5Ta-(0.06-0.70)O Alloys," *Ti-2003 Science and Technology Proceedings of the 10th World Conference on Titanium*, Wiley-VCH Verlag GmbH & Co., Weinheim, Germany, 2004.

[28] Qazi, J. I., et al., "Effect of Aging Treatments on the Tensile Properties of Ti-35Nb-5Zr-5Ta-(0.06-0.70)O Alloys," *Titanium, Niobium, Zirconium and Tantalum for Medical and Surgical Applications*, Washington, D.C., 2004.

Journal of ASTM International, July/August 2005, Vol. 2, No. 7
Paper ID JAI12775
Available online at www.astm.org

Gordon Hunter,[1] *Jim Dickinson,*[1] *Brett Herb,*[2] *and Ron Graham*[2]

Creation of Oxidized Zirconium Orthopaedic Implants

ABSTRACT: More demanding performance expectations for total joint arthroplasty are driving the development of alternative bearing materials. Oxidized zirconium was developed as an alternative to cobalt-chromium alloy for knee and hip femoral components in order to reduce wear of the polyethylene counterface and to address the needs of metal-sensitive patients. Oxidation in high temperature air transforms the metallic Zr-2.5Nb alloy surface into a stable, durable, low-friction oxide ceramic without creating the risk for brittle fracture associated with monolithic ceramic components. This presentation reviews aspects of this technology with a historical perspective, including standards for the zirconium alloy, non-medical applications for oxidized zirconium, and previous orthopaedic applications for zirconium. Manufacturing processes for oxidized zirconium components are described, beginning with refining of the zirconium from beach sand, to producing the alloy ingot and bar, to fabricating the component shape, and finally to oxidizing the surface and burnishing it to a smooth finish. Conditions are described for producing the oxide with excellent integrity, which is nominally 5 μm thick and predominantly monoclinic phase. The metal and oxide microstructures are characterized and related to the mechanical properties of the components and durability of the oxide. Laboratory hip and knee simulator tests are reviewed, which indicate that oxidized zirconium components reduce wear of the polyethylene counterface by 40–90 % depending on test conditions. As evidenced by promising early clinical experience, oxidized zirconium components have characteristics that provide an alternative to conventional cobalt-chromium components with an interchangeable surgical technique, while providing the potential for superior performance.

KEYWORDS: zirconium, zirconia, oxidation, manufacturing processes, alternative bearing materials, joint replacement, arthroplasty

Introduction

Wear of polyethylene bearing components is often cited as a reason for joint replacement complications and failure. Clinically retrieved cobalt-chromium hip and knee components exhibit roughening of their articular surfaces, and these features can increase wear of the opposing ultra-high molecular weight polyethylene components [1,2,3,4]. Despite improvements in the implant designs and quality of the polyethylene material, concerns remain about the adhesive and abrasive wear of the polyethylene as the metal surface roughens. Oxide ceramics, such as alumina and zirconia, have demonstrated improved wear performance over metal surfaces in numerous investigations [5]. The use of monolithic ceramics has been restricted because their brittle nature limits implant designs and creates concerns for component fracture. The desired alternative would combine the fracture toughness of metals with the wear performance of ceramics.

Oxidized zirconium was developed for orthopaedic applications to address these desires and provide improvements over cobalt-chromium alloy for resistance to roughening, frictional

Manuscript received 3 November 2004; accepted for publication 17 December 2004; published July 2005.
Presented at ASTM Symposium on Titanium, Niobium, Zirconium, and Tantalum for Medical and Surgical Applications on 9-10 November 2004 in Washington, DC.
[1] Senior Manager and Engineer, respectively, Smith & Nephew, Inc., Memphis, TN 38116.
[2] Manager and Director, respectively, ATI - Wah Chang, Albany, OR 97321.

behavior, and biocompatibility [6,7]. Prosthetic components are produced from a wrought zirconium alloy (Zr-2.5Nb) that is oxidized by thermal diffusion in heated air to create a cohesive and adherent zirconia surface. A small amount of niobium is alloyed with the zirconium metal to create a two-phase microstructure with sufficient strength and other mechanical properties for use as an orthopaedic prosthesis. The oxide is thick enough to provide the desired tribological properties against polyethylene or cartilage, but is thin enough that the component retains its metallic toughness and resilience.

Although similar in manufacture and properties to titanium, zirconium exhibits several unusual processing attributes, including its oxidation behavior, and has a unique commercial history. The manufacturing process begins with zirconium-containing beach sand and ends with an oxidized prosthesis (Fig. 1). Standardization efforts have been initiated for medical applications, while the successful clinical use of oxidized zirconium orthopaedic components continues to expand.

FIG. 1—*Zircon sand (left) is the raw material used to make an oxidized zirconium implant (right).*

Applications

Zirconium was discovered by Martin Klaproth in 1789 and was isolated as a metal by Jons Berzelius in 1824 [8]. It became available in commercial quantities only after the iodide decomposition process was developed in 1925, and then in greater quantities after the Kroll reduction technique was developed in 1947. At least one center in England used a metallic zirconium alloy (Zircaloy) for orthopaedic screws and bone plates in approximately 1950 [9]. The results were promising, but zirconium was made unavailable soon after that due to demand from government-sponsored nuclear power programs. Zirconium became available again outside these programs around 1958. By this time, titanium and its alloys had become less costly and more established clinically, so there was little interest in medical applications for zirconium alloys. In contrast, the chemical process industry became interested in using zirconium alloys for severe corrosion environments.

One of the attractive features of zirconium is the hard, adherent passive oxide film that it exhibits [8]. This film is autogenous, impervious, and hard. As a result, zirconium is used in corrosive applications where the film provides protection for the base metal in a wide range of aqueous acidic and basic solutions. The value of zirconium in the chemical process industry arises from its resistance to corrosion by the common inorganic and organic acids and their salts (except for hydrofluoric acid), and its almost complete inertness to inorganic and organic bases, either in aqueous solutions or as melts. Zirconium is used in very aggressive service including, for example, sulfuric acid up to the boiling point at 70 % acid concentration, to all concentrations of hydrochloric acid to well above the boiling point, and to concentrated (98 %) nitric acid up to near the boiling point [10].

The zirconium oxide film is generally self-healing in oxidizing media, and when grown thicker as a black oxide surface layer, it is extremely adherent and wear resistant [8,11]. Zirconium's low corrosion rate in the presence of superheated steam or supercritical boiling water is one of the properties that make it attractive for use as the material of choice for the structural support of the fuel cores for nuclear power plants. For nuclear energy, two other properties, low capture cross-section for neutrons (it is relatively transparent to neutrons and does not absorb them) and reasonable mechanical strength at about 300°C completes the suite of properties that make it the dominant core structural material for light water reactors [12].

This same surface oxide layer is utilized in erosion-corrosion applications where a combination of corrosion resistance and wear resistance is required from abrasive slurries of acidic media. Oxidation processes have been developed by artificially enhancing the autogenous oxide film, which can be done in a variety of ways [13,14]. Typical parts made with this treatment include ball valves, tee-joints, and elbows in piping systems. Hardness up to an equivalent of Rockwell 74C has been formed with the black oxide. An important application in this condition is for valves and pumps to handle debris-laden crude oil.

An aspect of zirconium's use in corrosion service is the low toxicity of the element and those of its compounds which contain no other toxic moiety [12,15]. At pH values associated with most biological tissues, zirconium moieties of most compounds are hydrolyzed to their oxides, the solubility of which is vanishingly small. Perhaps for this reason, most of the compounds do not exhibit any toxicity until the dosages are quite large. There are no reports of either the metal or its alloys with non-toxic constituents causing physiological reaction; indeed zirconium is a constituent of several alloys being developed for use in bone and joint prostheses [16,17]. Two of these alloys, known as Ti-13Nb-13Zr and Ti-12Mo-6Zr-2Fe, are the subjects of ASTM specifications F 1713 and F 1813, respectively.

There are several zirconium alloys utilized in the nuclear industry that could be considered for medical applications including the alloys known as Zircaloy-2, Zircaloy-4, Zr-2.5Nb, and Zr-1Nb. The first three of these alloys are covered under the Uniform Numbering System (UNS) designations R60802, R60804, and R60901, respectively. A second UNS designation for Zr-2.5Nb, R60904, was introduced in 1995 that increased the lower limit of the niobium concentration, and reduced the maximum allowable hydrogen from 25 to 10 ppm. The Zr-2.5Nb alloy has been used extensively for the manufacture of Canadian Deuterium Uranium (CANDU) reactor pressure tubes [11]. The niobium imparts creep resistance and additional elevated temperature strength. Niobium, in and of itself, is highly corrosion resistant, in addition to being a strengthener to zirconium. The Zr-1Nb alloy is being used as an advanced cladding material for French pressurized water reactors under the designation, M5, and for similar application in Russian reactors under the designation E110.

A slight interest in zirconium for medical applications, particularly in dental implants, has continued over the years because of its excellent biocompatibility [9,18,19]. In the late 1980s, research began to focus on growing the black surface oxide so that oxidized zirconium could be used in articulating applications for orthopaedic implants [6]. A properly produced oxide was found to have the attributes of low-friction, abrasion-resistance, and excellent adherence to the metal. Yet the material retained the toughness of the metal without the brittle behavior characteristic of monolithic zirconia ceramics. This led to the development and commercial release of tough, durable, and biocompatible oxidized zirconium bearing components for joint arthroplasty.

Standards

The Zr-2.5Nb alloy used for knee and hip prostheses was originally derived from chemical grade ZIRCADYNE™ 705 (ATI - Wah Chang, Albany, OR) under the UNS designation R60705. In order to gain possible advantages in improved oxidation behavior and reduced tissue exposure to sensitizing metal ions, development shifted to the tighter residual element controls of the specifications used for production of the CANDU pressure tubes under the UNS designation R60901, and later, R60904. The important ASTM specifications for this material include B 349 for sponge, B 350 for ingot, B 351 for rod and wire, and B 353 for seamless and welded tubes.

The development efforts focused on a commercial alloy to facilitate material production and inventory control. The manufacturing practices were well established at the time, and there was a substantial database of Zr-2.5Nb performance, particularly with respect to mechanical properties, fracture toughness, and oxide film formation and growth mechanisms [8,11]. Moreover, internationally recognized ASTM standards existed for this alloy. There was a substantial amount of Zr-2.5Nb being produced for CANDU pressure tubes, and this material stream provided a ready resource for the small quantities required for prosthetic development. There were a lot of early trials involving thermomechanical processing, including hot working variations, heat treatment variations, and chemistry variations to provide optimum microstructures for oxide formation, strength, and toughness. Because of the routine production of Zr-2.5Nb for nuclear applications,and the ready availability of material, the time to complete these trials and experiments could be condensed into a much shorter time than would normally be required. This work provided a firm baseline for the medical device production, and it became relatively easy to "piggy-back" onto the existing high volume product line with the early-on lower volume of medical device materials.

As the manufacturing processes for medical devices matured, work on a medical grade ASTM "F" specification for Zr-2.5Nb began in the fall of 2002. After several ballots and revisions, the document was approved as a standard by ASTM's Main Committee in the spring of 2004 and is being released as ASTM F 2384. Document construction relied heavily on the material specification "templates" that are continually updated and revised with the intent to keep ASTM "F" material specifications consistent.

The chemical requirements as listed in ASTM F 2384 were obtained from ASTM specification B 351 (UNS R60901). However, the list of elements was shortened to include only those required by ASTM specification B 550 (UNS R60705). The list of elements was shortened because meticulous control of residual elements is not needed for medical applications. The chosen compositional limits dictated the need to use the minimum mechanical property requirements for annealed material as listed in ASTM specification B 351 (UNS R60901).

Fabrication

Sponge

The production of Zr-alloys begins by reducing zirconium from what is known as zircon "sand" (Fig. 2). Zircon is the mineral form, zirconium orthosilicate, or $(Zr,Hf)SiO_4$. The principle sources of this material are from mining beach and dune deposits from Western Australia or the southeastern coast of the United States (US). The principal use of zircon is for ceramic glaze opacifiers, refractories, and metal casting molds. A small amount goes into glasses, advanced ceramics, abrasives, and zirconium chemicals. The use of zircon for reduction to the metal only comprises about 1 % of the total volume of zircon mined in the world. Estimates of crustal abundance place the value in the 100–200 ppm range. This means that zirconium is about as abundant as relatively common metals like copper, zinc, nickel, and chromium, and substantially more abundant than other well-known elements like tin, lead, bismuth, cadmium, antimony, and mercury [20].

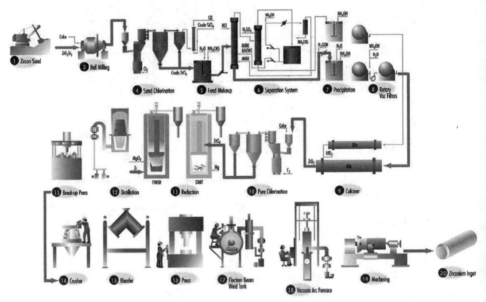

FIG. 2—*Zircon sand is refined to sponge and then alloyed to produce a Zr-alloy ingot.*

Zircon deposits mined in the US seldom contain over 5 % heavy minerals, of which only one-tenth or less is zircon. The remainder included ilmenite (a titanium mineral), monazite (a source of thorium and the lanthanides), garnet, and rutile and leucoxene (two titanium minerals). The sand is screened to remove debris and then subjected to gravity separation treatment with Reichert cones, spirals, and tray separators to remove the quartz and other low density minerals. The wet concentrate may be subjected to magnetic and electrostatic treatments to segregate the heavy minerals into separate components.

The first step in the decomposition of zircon, $ZrSiO_4$, is to remove the silica fraction. This is typically accomplished by carbochlorination in fluidized bed chlorinators that are externally heated. Zircon can be reacted with carbon, as petroleum coke, and chlorine to give zirconium tetrachloride and silicon tetrachloride at 1375 K according to the equation:

$$ZrSiO_4 + 4C + 4Cl_2 \rightarrow ZrCl_4 + SiCl_4 + 4CO \qquad (1)$$

Zirconium tetrachloride has a sublimation temperature of about 600 K and condenses from the chlorinator product gas as a loosely packed solid. It is collected in water-cooled condensers maintained above the condensation temperature of the silicon tetrachloride, which has a boiling point of about 330 K. The $SiCl_4$ is subsequently condensed as a liquid, and sold as a byproduct.

There are several processes used to separate the zirconium and hafnium including fractional crystallization, extractive distillation, and solvent extraction. Solvent extraction is the process practiced in the US. In this process, zirconium tetrachloride is hydrolyzed to zirconyl chloride solution, the acidity is adjusted with ammonia, and methyl isobutyl ketone is used to strip ferric ions. Then, ammonium thiocyanate is added to form zirconium and hafnium thiocyanate complexes, and the solution is contacted in continuous, countercurrent packed columns with methyl isobutyl ketone containing a small amount of thiocyanic acid. The hafnium is extracted to the less dense organic phase, and the zirconium segregates to the aqueous phase. This stream is scrubbed with pure ketone to remove the thiocyanic acid and treated with ammonium sulfate to precipitate a soluble sulfate. The zirconium sulfate is roasted to produce a pure anhydrous oxide.

The next steps in the operation are for the reduction to the metal. The calcined oxide is once again chlorinated much like the initial, or "sand chlorination" to further reduce impurities like aluminum, iron, and titanium. This process is known, appropriately enough, as pure chlorination. The resulting zirconium tetrachloride is reduced in a process named after its inventor, William Kroll. The Kroll reaction utilizes liquid magnesium as the reductant. The equation which governs this reaction at approximately 1000 K is:

$$2Mg + ZrCl_4 \rightarrow Zr + 2MgCl_2 \tag{2}$$

The magnesium is loaded into the bottom of the reduction vessel, or crucible. The top part of the retort comprises a cylindrical annulus filled with zirconium tetrachloride. In practice, the magnesium is melted first in the crucible, and then heat is applied to the upper retort to sublime the tetrachloride. The tetrachloride runs down a tubular "finger" and reacts with the molten magnesium to form a "sponge donut" in the lower crucible.

The zirconium, being denser, sinks to the bottom and sinters into a friable mass, termed "sponge." The magnesium chloride rises to the top and crystallizes into something that has the appearance of dirty snow. The sponge donut is stripped from the crucible. The magnesium chloride is mechanically removed as best as can be accomplished. The donuts are then placed in a vacuum atmosphere distillation furnace where the $MgCl_2$ is first melted out of the interstices of the sponge, and then evaporated as the temperature is raised to about 1250 K.

The distilled sponge is then broken up into large chunks and crushed to less than 25 mm particle size. Each donut is kept separated during this crushing, and a small fraction is sampled and melted into a "button." The button is used to chemically characterize the impurities in the sponge, the major ones being Hf, Al, Fe, C, O, Si, Ti, and P. Several donuts are blended together, based on impurity content, to provide a mass of roughly 8 MT. A second sample is obtained from the blend and used to make a small evaluation ingot using vacuum arc remelting (VAR). Samples from this ingot are again analyzed for impurity elements to confirm the blend results, and these chemistry values are then used as input to determine tolerable levels of recycle additions for subsequent VAR operations, along with alloy additions.

Ingot

Zirconium sponge, recycle material, and alloying elements are assembled into electrodes for consumable electrode VAR. Recycle material may be combined as chips or solids into the electrode. The recycle must be carefully cleaned, sampled, and analyzed to avoid unwanted impurities that can cause deviations in chemistry. Alloying elements, principally niobium and zirconium-oxide, are added to zirconium sponge during compaction into dense cylindrical billets on a hydraulic press. Individual billets of sponge and recycle are equally disbursed along the length and joined by electron beam welding to form a continuous electrode that can be 250–400 mm diameter and weigh up to 1.7 mt. The Zr-2.5Nb electrodes are melted multiple times into cylindrical ingots. After the initial melt, the electrodes can be combined as the ingot diameter increases during the subsequent melts to yield a final ingot weighing over 8 mt. Multiple VAR melts consolidate, refine, and homogenize the sponge, recycle, and alloy elements.

The consumable electrode is melted in a vacuum to prevent adsorption of oxygen and nitrogen from the atmosphere. Direct current power applied to the consumable electrode generates an electric arc to the starting base or molten pool. The arc locally melts the electrode on the end face, resulting in droplets falling down to form a molten pool. The molten zirconium pool is contained and solidified in a water-cooled copper crucible. The cast ingot is sampled at multiple positions along the length of the ingot to determine the composition, since the ingot is not completely molten at one time. The ASTM specification B 350 requires the first sample to be taken near the top of the ingot with additional samples taken at intervals equal to the ingot diameter down the bottom. The ASTM standard recognizes that the molten pool is at least one diameter deep at any time. Segregation is not a problem with current practice given the modern melt techniques and equipment.

Bar

The Zr-2.5Nb alloy forms a two-phase microstructure of hexagonal close-packed α-Zr and body-centered cubic β-Zr [8]. The ingot is heated to the β-phase and reduced through a sequence of steps by press forging. The purpose of press forging is twofold: first, it is used to reduce the physical dimensions of the as-cast ingot; second, it refines the ingot metallurgical grain structure. The as-cast ingot has a non-uniform grain structure consisting of long columnar grains and finer equiaxed grains. Preheating to the β-phase homogenizes the as-cast structure, and the strain energy from the forging allows for recrystallization to produce a final product with a uniform, fine grain structure. A second forging operation is usually performed in the α+β phase to the final billet size ranging from 65–215 mm diameter. The cooler working temperatures during forging impart additional strain energy for recrystallization.

The forging may receive a β-solution anneal to ensure homogeneous distribution of solid solution elements like oxygen and secondary phase-forming elements like niobium, iron, carbon, and silicon. The solution anneal is followed with water quenching. A homogeneous distribution of precipitated niobium-rich β-Zr particles is important for the formation of a uniform oxide layer in the final component.

The conversion of the intermediate-size billet to final-size bar can be accomplished by various hot and cold working processes. The hot working processes are performed in α+β phase field. Typically, processes include swaging, extrusion, bar rolling, and drawing to reduce the diameter to less than 65 mm diameter. All steps require stringent controls to minimize surface damage and impart the correct structure to the material. Zirconium tends to gall badly during working processes like extrusion or drawing. Development of lubrication packages minimizes the sticking of metal to tooling that produces a poor surface quality.

The bar can be conditioned by various methods to improve the surface quality. Typical steps include machining, grinding, belt sanding, sand blasting, and etching with a solution of nitric acid, hydrofluoric acid, and water.

Component

Having a small grain size in the metal is critical for producing good oxide integrity. Wrought bar with a refined microstructure can meet this grain size requirement. A bar of a sufficient diameter can be milled directly into the desired component shape. In order to reduce milling time and machining chips, close-die forging is often used to hot-work the bar closer to the final shape. Care must be taken to minimize the forging temperature in order to prevent excessive grain coarsening while the metal is heated. Contour-grinding also can be used to achieve tight dimensional tolerances, but surface discoloration by localized over-heating must be avoided for proper oxide formation.

Once the shape has been established, the component is oxidized by "baking" in air at a temperature above 500°C to allow oxygen to diffuse into the surface (Fig. 3). The oxide is not an externally applied coating, but rather a transformation of the original metal surface into zirconium-oxide ceramic after it becomes saturated with oxygen. Because it is controlled by a diffusion process, the oxide grows uniformly on all surfaces of even a complex shape. A diffusional gradient in oxygen enrichment is created under the oxide that provides a gradual transition in mechanical properties from the oxide to the metal. Unlike the abrupt change in strain behavior characteristic of overlay coatings, this transition zone enhances oxide adherence when the component is mechanically strained.

FIG. 3—*As oxygen diffuses into the heated surface, the metal transforms into a ceramic oxide (outer surface is at top in this cross-sectional metallograph that is heat-tinted to show the refined wrought microstructure).*

The desired oxide cohesiveness is created by the proper combination of many factors at the original metal surface. Castings have a coarse microstructure that allows non-uniform oxide growth, which can lead to crack formation within the oxide (Fig. 4). During oxidation, very smooth or contaminated surfaces can produce aesthetic and functional flaws in the oxide. An appropriate selection of time and temperature during oxidation can produce an oxide of uniform thickness without internal flaws. Continuing oxidation at a given temperature eventually leads to a transition in oxidation behavior during which the oxide starts cracking and growing rapidly and non-uniformly. Therefore, clean and moderately rough articular surfaces on fine-grained wrought material are oxidized at a tightly controlled time and temperature in order to produce a high integrity oxide of consistent thickness.

FIG. 4—*This centrifugal-casting created a non-uniform oxide with internal cracks (cross-sectional metallograph that is heat-tinted to show the coarse cast microstructure).*

A dense, fine-grained oxide approximately 5 μm thick is produced on OXINIUM™ oxidized zirconium components (Smith & Nephew, Inc., Memphis, TN). The oxidized component is then burnished to produce an articular surface at least as smooth as that of a cobalt-chromium component. The burnishing process allows polishing of complex shapes with a consistent removal of up to approximately 1 μm of oxide thickness. Because bone cement does not adhere well to the oxide, fixation surfaces are blasted with alumina grit to achieve adequate cement fixation. The finished result is a component with a configuration that is interchangeable with cobalt-chromium devices but with many superior performance attributes.

Performance

The zirconium alloy currently used to make oxidized zirconium prostheses contains a small amount of niobium and oxygen for mechanical strength. Like titanium, the metallic elements of zirconium and niobium are very biocompatible with minimum biological availability and electrocatalytic activity due to their passive oxide layers with extremely low solubility and excellent protective ability [16]. The biocompatibility of Zr-2.5Nb alloy both with and without an oxide was found to be at least equivalent to that of cobalt-chromium and Ti-6Al-4V alloys [7]. In addition, there are possible advantages for patients with metal allergies both because the alloy contains nearly undetectable levels of residual elements, including aggressive sensitizers such as nickel, cobalt, and chromium, and because the oxide inhibits metal ion release from the articular surfaces [5,21,22].

Mechanical testing and finite element analysis indicated that knee femoral components made from oxidized zirconium and cobalt-chromium alloy have equivalent device fatigue strength, with both exceeding a strength of 450 MPa for ten million cycles [23]. In contrast to the limitations exhibited by monolithic ceramics such as alumina and zirconia used in hip prostheses, oxidized zirconium modular heads do not exhibit brittle fracture during crush tests [24]. Because of the strength, toughness, and stability of oxidized zirconium, it can be used in components that duplicate the design of cobalt-chromium and other metal parts. This allows the flexibility of offering the advantages of ceramic components interchangeably with metallic components of the same design and without the requirement of unique surgical procedures and associated training.

The wrought metallic microstructure transforms during oxidation to a fine-grained (almost nanostructural) oxide that is predominantly monoclinic zirconia [25]. This crystal structure is fully stable, unlike the metastable tetragonal crystal structure of yttria-stabilized zirconia components [24]. Individual grains in the microstructure of stabilized zirconia have a tendency to transform to the monoclinic structure under certain conditions, which results in reduced strength and increased surface roughness. This does not happen to the monoclinic oxide on oxidized zirconium. The oxide is comprised primarily of staggered columnar grains oriented perpendicular to the outer surface. It is left in a compressive stress state without pores or voids internally or at the interface. The fine-grained structure provides some effective ductility so that the oxide can deform modestly without cracking. All of these microstructural features inhibit crack propagation and intra-oxide spallation under shear, contributing to the excellent integrity observed for this material.

The hardness of oxidized zirconium is over twice that of cobalt-chromium alloy on the articular surface, and the oxide is durable during articulation so that it has superior abrasion resistance [26,27]. To simulate a severe instance of third-body debris embedded in the polyethylene rubbing against the counterface for approximately ten years, bone cement pins were rubbed against flat plates for ten million cycles [28]. Oxidized zirconium plates exhibited over 4900 times less volumetric wear loss and over 160 times less roughness than cobalt-chromium alloy plates in this test. The oxidized surface did not chip off or wear away even when a strip of oxide was removed perpendicularly across the wear track before testing [29]. Not only did this demonstrate that oxidized zirconium is much more abrasion resistant than cobalt-chromium, but it also indicated that the oxide could tolerate localized damage or loss without catastrophic failure. Ceramic coatings and other surface modifications for orthopaedic applications have not demonstrated this level of durability [30,31].

The oxide surface can provide several advantages over cobalt-chromium for reducing wear of the polyethylene counterface, hereby reducing the generation of the sub-micrometer wear particles that are of concern for periprosthetic osteolysis [5]. Abrasive wear is caused as hard counterface asperities scrape against the polyethylene surface. Adhesive wear is caused by frictional shear as the polyethylene surface articulates against the hard counterface. Laboratory studies have shown that the oxide exhibits better wetting behavior in lubricating fluids and produces less friction than cobalt-chromium [32,33,34]. Thus, improving lubrication and maintaining smooth counterfaces can reduce small particle generation from these two wear mechanisms. These attributes also may provide similar benefits for reducing wear during direct articulation against cartilage [35].

Knee and hip simulator and other tests have been used to help understand how oxidized zirconium counterfaces compare to the standard cobalt-chromium counterfaces for wear of the polyethylene components [36,37,38]. Collectively, the simulator tests have indicated that oxidized zirconium components can reduce volumetric wear rates of the polyethylene counterface by 40–90 % depending on test conditions. The relative advantage for oxidized zirconium has been found to increase when testing with more demanding knee kinematics such as increased rotation and varus moment [39]. The relative advantage also increased when simulating more abrasive conditions for both knees and hips in comparison to testing with smooth components and filtered lubricant [40,41,42,43]. In comparison to conventional materials in hip tests, the combination of oxidized zirconium heads and highly crosslinked polyethylene liners reduced volumetric wear rates by 98 % and produced 81 % fewer debris particles [44]. Hip tests also showed that polyethylene remains cooler articulating against oxidized zirconium than either zirconia ceramic or cobalt-chromium, and that oxide damage does not lead to a catastrophic implant failure [45,46]. These results indicate that oxidized zirconium may contribute to reducing wear-related clinical complications such as debris-induced osteolysis, particularly for "high demand" patients that are most in need of assistance for prolonging the survival of their prostheses.

Over 50 000 total knee replacements have been performed using oxidized zirconium femoral components since the first surgery in 1997 through the middle of 2004. Over 15 000 total hip replacements have been performed using oxidized zirconium femoral heads since the first surgery in 2002. Numerous clinical studies are being organized to compare the performance of oxidized zirconium and cobalt-chromium components. The first study to be published involved knee patients, and no adverse effects had been observed at the two-year evaluation [47]. In the randomized prospective portion of this study, the oxidized zirconium patients experienced a more rapid return of flexion and regaining of functional milestones. A much longer time period will be required to measure differences in polyethylene wear and implant survivorship.

Summary

Zirconium alloys have been in commercial production for many decades and are used successfully in several demanding applications. Oxidation transforms the metallic surface into a stable, durable, low-friction surface that has several potential advantages for load-bearing articular components in orthopaedic applications. Clinical use of oxidized Zr-2.5Nb continues to expand, and early results appear promising. As a result, ASTM F 2384 has been established for medical applications of this zirconium alloy.

Acknowledgements

The metallographic contributions of Carolyn Weaver are appreciated.

References

[1] Barrack, R. L., Castro, Jr., F. P., Szuszczewicz, E. S., and Schmalzried, T. P., "Analysis of Retrieved Uncemented Porous-Coated Acetabular Components in Patients with and without Pelvic Osteolysis," *Orthopedics*, Vol. 25, No. 12, 2002, pp. 1373–1378.
[2] Levesque, M., Livingston, B. J., Jones, W. M., and Spector, M., "Scratches on Condyles in Normal Functioning Total Knee Arthroplasty," *Transactions of the Orthopaedic Research Society*, Vol. 23, 1998, p. 247.
[3] Fisher, J., Firkins, P., Reeves, E. A., Hailey, J. L., and Isaac, G. H., "The Influence of Scratches to Metallic Counterfaces on the Wear of Ultra-High Molecular Weight Polyethylene," *Proceedings of the Institute of Mechanical Engineers*, Vol. 209-H, 1995, pp. 263–264.
[4] Dowson, D., Taheri, S., and Wallbridge, N. C., "The Role of Counterface Imperfections in the Wear of Polyethylene," *Wear*, Vol. 119, 1987, pp. 277–293.
[5] Davidson, J. A., "Characteristics of Metal and Ceramic Total Hip Bearing Surfaces and Their Effect on Long-Term Ultra High Molecular Weight Polyethylene Wear," *Clinical Orthopaedics and Related Research*, Vol. 294, 1993, pp. 361–378.
[6] Davidson, J. A., Asgian, C. M., Mishra, A. K., and Kovacs, P., "Zirconia (ZrO$_2$)-Coated Zirconium-2.5Nb Alloy for Prosthetic Knee Bearing Applications," *Bioceramics Volume 5*, Kobunshi Kankokai, T. Yamamuro, T. Kokubo, and T. Nakamura, Eds., Kyoto, 1992, pp. 389–401.
[7] Hunter, G., Jones, W. M., and Spector, M., "Oxidized Zirconium," *Total Knee Arthroplasty*, Springer-Verlag, J. Bellemans, M. D. Ries, and J. Victor, Eds., in press.
[8] Schemel, J. H., *ASTM Manual on Zirconium and Hafnium, STP 639*, ASTM International, West Conshohocken, PA, 1977.
[9] Laing, P. G., "Historical Perspective on Titanium Implants," ASTM Symposium on Medical

Applications of Titanium and Its Alloys, Phoenix, 15–16 November 1994.

[10] Yau, T. L. and Webster, R. T., "Corrosion of Zirconium and Hafnium," *Metals Handbook, Volume 13, Ninth Edition*, American Society for Metals, Metals Park, OH, 1987, pp.707–721.

[11] Cox, B., "Oxidation of Zirconium and Its Alloys," *Advances in Corrosion Science and Technology, Volume 5*, Plenum Press, M. G. Fontana and R. M. Staehle, Eds., New York, 1976, pp. 173–391.

[12] Haygarth, J. C. and Graham, R. A., "Zirconium and Hafnium," *Review of Extraction, Processing, Properties & Applications of Reactive Metals*, The Minerals, Metals & Materials Society, B. Mishra, Ed., Warrendale, PA, 2001, pp. 1–72.

[13] Haygarth, J. C. and Fenwick, L. J., "Improved Wear Resistance of Zirconium by Enhanced Solid Oxide Films," *Thin Solid Films*, Vol. 118, 1984, pp. 351–362.

[14] Kemp, W. E., "Nobleizing: Creating Tough, Wear Resistant Surfaces on Zirconium," *Outlook*, Vol. 11, No. 2, 1990, pp. 4–8.

[15] Blumenthal, W. B., "Zirconium - Behavior in Biological Systems," *Journal of Scientific and Industrial Research*, Vol. 35, No. 7, 1976, pp. 485–490.

[16] Kovacs, P. and Davidson, J. A., "Chemical and Electrochemical Aspects of the Biocompatibiltiy of Titanium and Its Alloys," *Medical Applications of Titanium and Its Alloys, STP 1272*, S. A. Brown and J. E. Lemons, Eds., ASTM International, West Conshohocken, PA, 1996, pp. 163–178.

[17] Davidson, J. A., Mishra, A. K., Kovacs, P., and Poggie, R. A., "New Surface-Hardened, Low-Modulus, Corrosion-Resistant Ti-13Nb-13Zr Alloy for Total Hip Arthroplasty," *Bio-Medical Materials and Engineering*, Vol. 4, No. 3, 1994, pp. 231–243.

[18]"Surgical Implants Made of Zirconium," N. A. Dollezhal Research and Development Institute of Power Engineering, Moscow, Russia, 4 August 2004. URL: http://www.nikiet.ru/eng/structure/hightemp/implant.html.

[19] "Implants," http://www.zahnweiss.at/e/implant.htm , C. S. Weiss, Vienna, Austria, 4 August 2004.

[20] Erlank, A. J., et al., *Handbook of Geochemistry, Volume II/4*, Springer-Verlag, K. H. Wedepohl, Ed., Heidelberg, Germany, 1978, section 72-E.

[21] Hallab, N., Merritt, K., and Jacobs, J. J., "Metal Sensitivity in Patients with Orthopaedic Implants," *The Journal of Bone and Joint Surgery*, Vol. 83-A, No. 3, 2001, pp. 428–436.

[22] Vittetoe, D. A. and Rubash, H. E., "Strategies for Reducing Ultra-High Molecular Weight Polyethylene Wear and Osteolysis in Total Knee Arthroplasty," *Seminars in Arthroplasty*, Vol. 13, No. 4, 2002, pp. 344–349.

[23] Tsai, S., Sprague, J., Hunter, G., Thomas, R., and Salehi, A., "Mechanical Testing and Finite Element Analysis of Oxidized Zirconium Femoral Components," *Transactions of the Society For Biomaterials*, Vol. 24, 2001, p. 163.

[24] Sprague, J., Aldinger, P., Tsai, S., Hunter, G., Thomas, R., and Salehi, A., "Mechanical Behavior of Zirconia, Alumina, and Oxidized Zirconium Modular Heads," *ISTA 2003 Volume 2*, International Society for Technology in Arthroplasty, S. Brown, I. C. Clarke, and A. Gustafson, Eds., Birmingham, AL, 2004, pp. 31–36.

[25] Benezra, V., Mangin, S., Treska, M., Spector, M., Hunter, G., and Hobbs, L. W., "Microstructural Investigation of the Oxide Scale on Zr-2.5Nb and Its Interface with the Alloy Substrate," *Biomedical Materials, Symposium Proceedings 550*, Materials Research Society, T. Neenan, M. Marcolongo, and R. F. Valentini, Eds., Warrendale, PA, 1999, pp. 337–342.

[26] Long, M., Riester, L., and Hunter, G., "Nano-Hardness Measurements of Oxidized Zr-2.5Nb and Various Orthopaedic Materials," *Transactions of the Society For Biomaterials*, Vol. 21, 1998, p. 528.

[27] Davidson, J. A., Poggie, R. A., and Mishra, A. K., "Abrasive Wear of Ceramic, Metal, and UHMWPE Bearing Surfaces from Third-Body Bone, PMMA Bone Cement and Titanium Debris," *Bio-Medical Materials and Engineering*, Vol. 4, No. 3, 1994, pp. 213–229.

[28] Hunter, G. and Long, M., "Abrasive Wear of Oxidized Zr-2.5Nb, CoCrMo, and Ti-6Al-4V

Against Bone Cement," *Sixth World Biomaterials Congress Transactions*, Society For Biomaterials, Minneapolis, 2000, p. 835.

[29] Hunter, G., "Adhesion Testing of Oxidized Zirconium," *Transactions of the Society For Biomaterials*, Vol. 24, 2001, p. 540.

[30] Hunter, G., Pawar, V., Salehi, A., and Long, M., "Abrasive Wear of Modified CoCr and Ti-6Al-4V Surfaces Against Bone Cement," *Medical Device Materials*, ASM International, S. Shrivastava, Ed., Materials Park, OH, 2004, pp. 91–97.

[31] Mishra, A. K. and Davidson, J. A., "Zirconia/Zirconium: A New, Abrasion Resistant Material for Orthopaedic Applications," *Materials Technology*, Vol. 8, Nos. 1/2, 1993, pp. 16–21.

[32] Mazzucco, D. and Spector, M., "Tribological Evaluation of Oxidized Zirconium Versus Cobalt-Chromium Alloy Against Polyethylene," *Transactions of the Orthopaedic Research Society*, Vol. 29, 2004, p. 1460.

[33] Salehi, A., Aldinger, P., Sprague, J., Hunter, G., Bateni, A., Tavana, H., et al., "Dynamic Contact Angle Measurements on Orthopaedic Ceramics and Metals," *Medical Device Materials*, ASM International, S. Shrivastava, Ed., Materials Park, OH, 2004, pp. 98–102.

[34] Poggie, R. A., Wert, J. J., Mishra, A. K., and Davidson, J. A., "Friction and Wear Characterization of UHMWPE in Reciprocating Sliding Contact with Co-Cr, Ti-6Al-4V and Zirconia Implant Bearing Surfaces," *Wear and Friction of Elastomers, STP 1145*, R. Denton and M. K. Keshavan, Eds., ASTM International, West Conshohocken, PA, 1992, pp. 65–81.

[35] Patel, A. M. and Spector, M., "Tribological Evaluation of Oxidized Zirconium Using an Articular Cartilage Counterface," *Biomaterials*, Vol. 18, No. 5, 1997, pp. 441–447.

[36] Spector, M., Ries, M. D., Bourne, R. B., Sauer, W. S., Long, M., and Hunter, G., "Wear Performance of Ultra-High Molecular Weight Polyethylene on Oxidized Zirconium Total Knee Femoral Components," *The Journal of Bone and Joint Surgery*, Vol. 83-A, Supplement 2, 2001, pp. 80–86.

[37] White, S. E., Whiteside, L. A., McCarthy, D. S., Anthony, M., and Poggie, R. A., "Simulated Knee Wear with Cobalt Chromium and Oxidized Zirconium Knee Femoral Components," *Clinical Orthopaedics and Related Research*, Vol. 309, 1994, pp. 176–184.

[38] Walker, P. S., Blunn, G. W., and Lilley, P. A., "Wear Testing of Materials and Surfaces for Total Knee Replacement," *Journal of Biomedical Materials Research*, Vol. 33, No. 3, 1996, pp. 159–175.

[39] Hermida, J. C., Patil, S., D'Lima, D. D., Colwell, Jr., C. W., and Ezzet, K. A., "Polyethylene Wear Against Metal-Ceramic Composite Femoral Components," *American Academy of Orthopaedic Surgeons Proceedings*, Vol. 5, 2004, p. 449.

[40] Ries, M., Salehi, A., Widding, K., and Hunter, G., "Polyethylene Wear Performance of Oxidized Zirconium and Cobalt-Chromium Knee Components Under Abrasive Conditions," *The Journal of Bone and Joint Surgery*, Vol. 84-A, Supplement 2, 2002, pp. 129–135.

[41] DesJardins, J. D. and LaBerge, M., "UHMWPE In-Vitro Wear Performance Under Roughened Oxidized Zirconium and CoCr Femoral Knee Components," *Transactions of the Society For Biomaterials*, Vol. 26, 2003, p. 364.

[42] Good, V., Ries, M., Barrack, R. L., Widding, K., Hunter, G., and Heuer, D., "Reduced Wear with Oxidized Zirconium Femoral Heads," *The Journal of Bone and Joint Surgery*, Vol. 85-A, Supplement 4, 2003, pp. 105–110.

[43] Clarke, I. C., Green, D. D., Williams, P. A., and Good, V., "Simulator Comparison of XLPE Wear with 36 mm CoCr and Oxidized Zirconium Balls in Smooth and Roughened Condition," *Transactions of the Seventh World Biomaterials Congress*, Australian Society for Biomaterials, Victoria, Australia, 2004, p. 1138.

[44] Good, V., Widding, K., Heuer, D., and Hunter, G., "Reduced Wear Using the Ceramic Surface on Oxidized Zirconium Heads," *Bioceramics in Joint Arthroplasty*, Steinkopff Verlag, J. Y. Lazennec and M. Dietrich, Eds., Darmstadt, Germany, 2004, pp. 93–98.

[45] Tsai, S., Aldinger, P., Hunter, G., Thornberry, R. L., and Salehi, A., "Thermal Generation and Dissipation Behavior of Various Bearing Materials in a Hip Joint Simulator,"

Transactions of the Seventh World Biomaterials Congress, Australian Society for Biomaterials, Victoria, Australia, 2004, p. 816.

[46] Heuer, D., Good, V., and Widding, K., "Wear Performance of Damaged Oxidized Zr-2.5Nb Modular Femoral Heads," *Transactions of the Society For Biomaterials*, Vol. 26, 2003, p. 366.

[47] Laskin, R. S., "An Oxidized Zr Ceramic Surfaced Femoral Component for Total Knee Arthroplasty," *Clinical Orthopaedics and Related Research*, Vol. 416, 2003, pp. 191–196.

Journal of ASTM International, November/December 2005, Vol. 2, No. 10
Paper ID JAI12777
Available online at www.astm.org

D. J. Medlin,[1] *J. Scrafton,*[1] *and R. Shetty*[2]

Metallurgical Attachment of a Porous Tantalum Foam to a Titanium Substrate for Orthopedic Applications

ABSTRACT: A porous tantalum foam (Trabecular Metal™) has been successfully attached to a titanium substrate for an orthopedic acetabular cup application by two different methods: sintered powder processing and diffusion bond processing. A metallurgical evaluation of the interfaces between the tantalum foam and the titanium substrates has been performed, as well as long-term corrosion immersion tests to evaluate the susceptibility of the materials and design to crevice corrosion attack. The sintered powder attachment method used sprayed titanium powder at the interface between the tantalum foam and the titanium substrate. The assembly is sintered at high temperature to form a metallurgical bond between the sintered titanium powder and the titanium substrate and a predominately mechanical interlock between the sintered powder and the tantalum foam. The diffusion bond attachment process used time, pressure, and temperature to form a metallurgical bond between the tantalum foam and the titanium substrate by solid state diffusion. Both methods formed acceptable bond strengths with no evidence of corrosion attack from the corrosion immersion tests.

KEYWORDS: tantalum foam, Trabecular Metal™, titanium alloy, corrosion, microstructure, orthopedic implants, diffusion bonding, sintering, titanium powder, acetabular cup

Introduction

A highly porous tantalum foam (Trabecular Metal™) has been developed for a variety of orthopedic applications with unique physical, mechanical, and tissue ingrowth characteristics [1–5]. The porosity of this biomaterial ranges from 75–85 % and is substantially higher than other porous coatings such as plasma spray, beads, and fiber mesh that have much lower porosities that range from 30–50 %. This biomaterial has been designed to withstand many physiological loads typical of human dynamics, however, some orthopedic applications require attaching this biomaterial to a metallic substrate, such as titanium, to withstand high impact and fatigue environments. Finding an attachment method that will form a metallurgical bond between the tantalum foam and the titanium substrate without any adverse corrosion reactions has benefits for many orthopedic implant applications.

Materials

Porous Tantalum Components

The porous tantalum foam manufacturing process begins with a medical grade polyurethane foam with a specific cell size between 400 and 600 μm and an aspect ratio of approximately 1.0. The polyurethane cell density and pore geometry dictate the final pore size and homogeneous

Manuscript received 19 October 2004; accepted for publication 25 April 2005; published November 2005.
Presented at ASTM Symposium on Titanium, Niobium, Zirconium, and Tantalum for Medical and Surgical Applications on 9-10 November 2004 in Washington, DC.
[1] Principal Engineer, Metals Research, Zimmer, Incorporated, P.O. Box 708, Warsaw, IN, 46581, USA.
[2] Director, Metals Research, Zimmer, Incorporated, P.O. Box 708, Warsaw, IN 46581, USA.

mechanical properties of the tantalum foam. The polyurethane foam is reticulated and then pyrolyzed, forming a low-density reticulated vitreous carbon foam (RCV). The RVC foam can then be machined or crushed into a variety of shapes and preforms. Next, the RVC preforms are coated with tantalum by a proprietary chemical vapor deposition/infiltration process (CVD/CVI) that deposits commercially pure tantalum throughout the RVC preforms. Increased amounts of deposited tantalum result in increased mechanical strength of the final tantalum foam product and a decreased metal foam porosity. The porosity (pore volume) of the final tantalum foam ranges from approximately 75–85 % depending upon the length of time in the CVD/CVI reactor and has an average pore size diameter between 400 and 600 μm [5].

The tantalum foam components evaluated in this study were manufactured at the Zimmer-TMT facility in Livingston, NJ. Two different sizes of tantalum foam acetabular shells were made for this analysis, 38 mm and 40 mm outside diameter shells, and the tantalum foam had a minimum average relative density of 18 %. Figure 1 shows an image of two representative acetabular shell components.

FIG. 1—*Tantalum foam acetabular shells 40 mm in diameter.*

Titanium Alloy Components

The titanium substrates that the tantalum foam shells were attached to were made from Ti-6Al-4V-ELI alloy that met the chemical composition and mechanical property requirements of ISO 5832-3 and ASTM Standard Specification for Wrought Titanium-6Aluminum-4Vanadium ELI (Extra Low Interstitial) Alloy for Surgical Implant Applications (ASTM F 136) [6]. The substrate acetabular cups were machined from bar stock and cleaned to remove any machining residue before bonding to the tantalum shells.

Titanium Powder

The titanium powder used for the sintered powder attachment method met the chemical composition requirements of ISO 5832-2 and ASTM Standard Specification for Unalloyed Titanium for Surgical Implant Applications (ASTM F 67) Grade-1 [6]. The particle size of the gas atomized spherical powder was characterized by sieve analysis to have a 95 % size distribution of –45 μm in diameter or finer.

Analytical Methods

Microstructural Analysis

Metallographic sections from bonded acetabular cups were sectioned in the radial (longitudinal) and circumferential (latitudinal) orientations and metallographically prepared with traditional techniques according to ASTM Standard Guide for Preparation of Metallographic Specimens (ASTM E 3) [7]. Metallographic samples were etched with a modified Kroll's Reagent for standard metallographic analysis and with Weck's Reagent for color metallographic analysis for enhanced resolution (see Table 1) [8–10]. Typical microstructural evaluations were performed with a Zeiss Axiovert 200 MAT metallograph with an AxioVision 4.2 image analysis system. Electron microscopy was performed with a Cambridge Strereoscan 360 with Tracor Nothern Z-Max 30 Series EDS system for elemental analysis.

TABLE 1—*The etchants used for the metallographic analysis.*
Weck's Reagent is a color etchant [8–10].

Etchant Name	Procedure	Composition
Kroll's Reagent	Immerse for 5–30 s	10 ml HF 5 ml HNO$_3$ 85 ml H$_2$O
Weck's Reagent	Immerse for 15–30 s	5 g ammonium bifluoride 4 ml HCl 100 ml H$_2$O

Corrosion Testing

Laboratory corrosion testing of the bonded tantalum foam and titanium substrate components was performed by soaking the specimens in a long-term corrosion immersion test. This test method evaluates the susceptibility of the metal component to crevice corrosion at the interface between the porous coating and substrate. Bonded specimens were ultrasonically cleaned in ethanol for 5 min, passivated in a 40 % nitric acid bath at room temperature for 20 min, rinsed in 3 successive water baths, and then hot air dried. The passivation process followed the ASTM Standard Practice for Surface Preparation and Marking of Metallic Surgical Implants (ASTM F 86) [6]. The specimens were then placed in a beaker with Ringer's Solution at a pH of 7.8, and air was slowly pumped (bubbled) into the beaker to keep the Ringer's Solution saturated with oxygen. The beakers were placed in a Lindberg/Blue-M constant temperature bath held at 37°C (±1°C) and monitored for a minimum of 6 months. After 6 months the samples were dried and evaluated for corrosion residue.

An accelerated corrosion immersion test was also performed on samples in which the same procedure was followed, except the pH of the Ringer's Solution was kept at 3.5 and the temperature of the bath was held at 50°C (±2°C). The purpose of this test was to enhance the conditions for crevice corrosion reactions at the interfaces of the samples. The samples were monitored for 3 months before removing and evaluating for corrosion residue.

Mechanical Testing

Tensile testing was performed on specimens after bonding to evaluate the strength of the attachment method. Three 12.7 mm diameter test cores were electrical discharge machined (EDM) from each bonded acetabular shell. The specimens were taken roughly halfway between the pole and equator of the hemisphere (cup). The titanium and tantalum foam ends of each electrical discharged core specimen were faced-off by lathe machining to make them as flat and parallel as possible. Material was removed from each surface to produce parallel flat surfaces on the core specimen, leaving a core specimen height of 5.0 mm (± 0.50 mm). Core specimens were assembled between two test blanks for tensile testing. The titanium and tantalum foam ends of the core specimen were adhered to the test blanks with 2–3 layers of FM-1000 epoxy film (Cytec, Havre de Grace, MD). Core specimens were tensile tested in an Instron 1125 universal test frame using an Instron Series IX software control system. The specimens were tested at a crosshead displacement rate of 2.54 mm per min.

Attachment Methods

Sintered Powder Method

The tantalum foam acetabular shells have a rough surface texture due to the open porosity and the protruding struts of the material. Figure 2 is an electron microscopy image of the tantalum foam showing the open porosity and the uneven surface texture inherent with this material. Direct diffusion bonding between the tantalum foam and the titanium substrate would be difficult due to the inability to maintain sufficient surface contact between the two metal surfaces without having to machine (smooth) the textured surface of the tantalum foam. To avoid machining the surface of the tantalum foam, a titanium powder mixed with a proprietary polymer binder was sprayed onto the inside diameter of the tantalum foam shell. The sprayed powder penetrated into the tantalum foam, and then a 1–2 mm depth of powder was subsequently built up with additional spray passes. The titanium cup substrate was then pressed into the inside diameter of the tantalum shell with an Arbor press. The titanium powder/binder interlayer filled the gap between the tantalum foam shell and the titanium cup substrate. Next, the acetabular cup assemblies were placed into a vacuum furnace, heated through a series of debind cycles, and sintered at approximately 1200°C for a few minutes [11]. The final product has a tantalum foam structure metallurgically attached to a titanium substrate.

Diffusion Bonding Method

The inside diameter of the tantalum foam shells were machined smooth on an end mill to reduce the surface topography variation and to increase the surface contact area between the tantalum foam shell and the titanium cup substrate. This eliminates the need to fill the space between the two metal surfaces. The foam shells and the substrate cups were assembled (without a powder interlayer), placed between two graphite fixtures, and the fixtures were bolted together compressing the tantalum foam shell and the titanium cup substrate. A combination of the smooth machined tantalum foam surface and the applied pressure from the carbon fixtures resulted in intimate contact between the foam and substrate surfaces. The entire assembly was vacuum heated treated at approximately 940°C for approximately 4 h. The combination of time, temperature, and pressure causes solid state diffusion transport between the tantalum and titanium metals and results in a metallurgical bond at the interface [11].

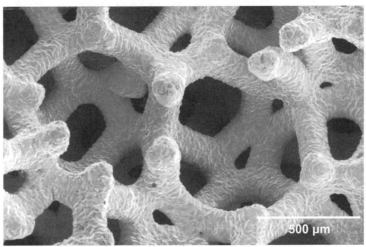

FIG. 2—*Electron microscopy image showing the surface of the tantalum foam.*

Results and Discussion

Microstructural Analysis of the Sintered Powder Method

The sintered cups were sectioned and photographed to show the composite structure. Figure 3 shows the porous tantalum foam, the sintered titanium powder interlayer, and the titanium substrate cup after processing. The arrows in Fig. 3 identify the sintered powder interlayer that bonds the tantalum foam to the titanium substrate cup and the complete fill of the titanium powder from the spray process and sintering heat treatment. The open-cell porous structure of the tantalum foam is shown in Fig. 4 at higher magnification. The sintered titanium powder layer is between the tantalum foam and the titanium substrate cup and is continuous and fills the entire gap at locations where it was intended to be sprayed. Furthermore, the sintered powder penetrates into the tantalum foam layer, forming a predominately mechanical interlock. It is the mechanical interlock between the sintered titanium powder and the open-cell porous foam that attaches (bonds) the tantalum foam to the sintered powder matrix. Figure 5 shows the sintered titanium powder penetration into the tantalum foam structure. The titanium powder penetration into the near surface of the tantalum foam is due to a combination of spray pressure and pressing the assembly together with an Arbor press during assembly.

The interface between the titanium substrate and the sintered titanium powder is shown in Fig. 6a. A continuous metallurgical bond is present between the titanium substrate and the sintered titanium powder when these two metals are in contact at these sintering temperatures. The titanium substrate microstructure is categorized as a Wildmanstatten alpha and beta titanium microstructure with acicular white plates of alpha phase and narrow dark areas of beta phase between the alpha plates. This type of microstructure is typical when a titanium alloy is slow cooled from above the beta transus temperature, as was done with these acetabular cups during the sintering process [12].

Figure 6*b* shows the interface between the surface of the tantalum foam and the sintered titanium powder. Only a very small portion of the total interface shows a distinct metallurgical bond between these two metals, and therefore, this particular interface is classified as a mechanical interlock, as previously described.

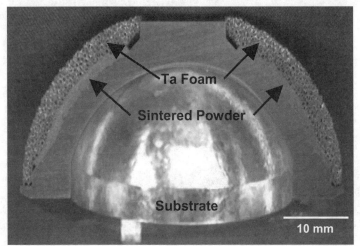

FIG. 3—*A sectioned acetabular cup manufactured with the sintered powder method.*

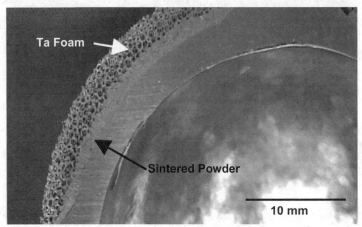

FIG. 4—*A sectioned acetabular cup showing the interface between the tantalum foam shell and the titanium substrate cup.*

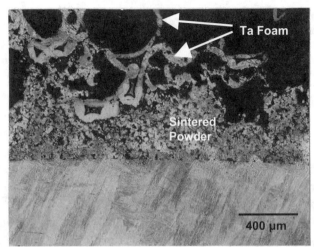

FIG. 5—*A light metallographic image showing the interface between the tantalum foam and the titanium substrate. Etched with Modifies Kroll's Reagent.*

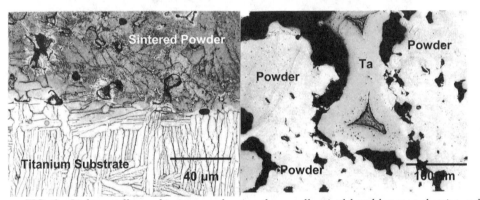

FIG. 6—*Light metallographic images showing the metallurgical bond between the sintered powder and the titanium substrate (a) and the interface between the tantalum foam and the sintered titanium powder (b). The left microstructure is etched with modified Kroll's Reagent, and the microstructure on the right is unetched.*

Figures 5, 6*a*, and 6*b* show evidence of the original spherical shape of the powder from before the sintering process. The commercially pure titanium powder was approximately 40 μm diameter spherical particles before spraying, and then the fine powder particles coalesced together during the sinter process cycle to form a large conglomerate mass between the tantalum foam shell and the titanium substrate. During sinter processing, the individual powder particles formed single crystals, which after cooling predominately appear as featureless single alpha phase crystals. The dark, or gray shadow, appearance of the sintered powder is due to slight over etching of the microstructure. In an effort to get the tantalum microstructure to etch and the fact that the porous structure of the sintered powder and tantalum foam will entrap acid during etching, the titanium powder unavoidably gets over etched during sample preparation and appears very dark [13].

Microstructural Analysis of the Diffusion Bond Method

Representative metallographic images showing the diffusion bonding interface between the porous tantalum foam and the titanium substrate are shown in Figs. 7 and 8. The tantalum metal is relatively passive to the chemicals in the Kroll's Reagent and, therefore, appears white in the images. The Kroll's Reagent reveals the grain structure of the titanium substrate and of the interface microstructure between the porous tantalum foam and the titanium (substrate).

FIG. 7—*Light micrographic images that show the interface between the tantalum foam (Ta) and the titanium substrate. Etched with Modified Kroll's Reagent.*

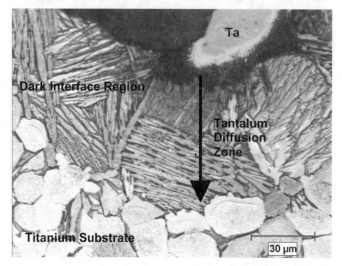

FIG. 8—*Light metallographic image showing the diffusion bonded interface between the porous tantalum foam (Ta) and the titanium substrate. The dark interface region reveals the depth of tantalum diffusion. Etched with Modified Kroll's Reagent.*

Examination of the microstructure between the tantalum foam and titanium components indicates an excellent solid state metallic diffusion bonded interface. The dark microstructure approximately 80–100 μm thick between the porous tantalum and titanium substrate shows a transformed beta microstructure attributed to the high beta stabilizing effect of elemental tantalum in titanium and titanium alloy systems [12]. The diffusion of tantalum into the titanium microstructure to a depth of almost 100 μm is in agreement with theoretical diffusion rate calculations and predictions associated with standard tantalum-titanium constitution (phase) diagrams. Furthermore, the tantalum-titanium phase diagram indicates that elemental tantalum and titanium have complete solid state solubility and will not form intermetallic compounds that could potentially embrittle the diffusion bonded interface [14]. The titanium substrate exhibits an equiaxed alpha-beta microstructure typical of Ti-6Al-4V-ELI alloys that have been heat treated in the two-phase alpha-beta region and slow cooled.

Mechanical Bond Strength

The tensile strength of specimens taken from cups that were processed with the sintered powder method had an average strength of 34.6 MPa (± 8.9), and the specimens taken from cups processed with the diffusion bond method had an average strength of 49.5 MPa (±7.5). Both of these interface tensile strengths exceed the 20 MPa minimum strength recommendation described in the FDA Guidance Document for Testing Orthopedic Implants with Modified Metallic Surfaces [15]. Most of the specimens fractured in the porous tantalum foam.

Corrosion Test Results

Specimens from both attachment processes were corrosion immersion tested for 6 months, and no evidence of corrosion residue was found on the samples. The samples were returned to the immersion test for an additional 18 months (24 months total) and, again, no evidence of corrosion residue was found.

Specimens that were immersion tested for 3 months in the accelerated corrosion test did not show any indications of corrosion activity and were returned to the accelerated test for a total of 12 months with no evidence corrosion residue.

Conclusions

Two successful methods have been developed to attach a tantalum foam to a titanium substrate for an acetabular cup component. The method of using a sprayed titanium powder as an interlayer between the tantalum foam and the titanium substrate resulted in a metallurgical bond between the sintered titanium powder and the substrate and a predominately mechanical interlock between the tantalum foam and the sintered powder. The diffusion bond method resulted in a metallurgical bond between the tantalum foam and the titanium substrate with evidence of tantalum diffusion into the titanium. Both attachment methods resulted in satisfactory mechanical bond strengths.

Both attachment methods did not show any evidence of crevice corrosion attack after immersion testing for 24 months in the 37°C and 7.8 pH test, and also no evidence was found after testing for 12 months in the 50°C and 3.5 pH test.

Both of these attachment methods may have application in other medical implant devices that have titanium substrates with a porous tantalum foam.

Acknowledgments

The authors would like to recognize the following colleagues for their valued assistance and support of this research and development project: C. Blanchard, M. Hawkins, S. Charlebois, D. Blakemore, D. Swarts, C. Panchison, and K. England.

References

[1]   Zardiackas, L. D., Parsell, D. E., Dillon, L. D., Mitchell, D. W., Nunnery L. A., and Poggie R., "Structure, Metallurgy and Mechanical Properties of a Porous Tantalum Foam," *J. Biomed Mater Res (Appl Biomater)*, Vol. 58, 2001, pp. 180–187.

[2]   Fitzpatrick, D., Ahn, P., Brown, T., and Poggie, R., "Friction Coefficients of a Porous Tantalum and Cancellous & Cortical Bone," *21st Annual American Society of Biomaterials*, Clemson, SC, September 24, 1997.

[3]   Bobyn, J. D., Stackpool, G. J., Hacking, S. A., Tanzer, M., and Krygier, J. J., "Characteristics of Bone Ingrowth and Interface Mechanics of a New Porous Tantalum Biomaterial," *J. Bone and Joint Surgery*, Vol. 81-B, No. 5, September 1999, pp. 907–914.

[4]   Hacking, S. A., Bobyn, J. D., Toh, K. K., Tanzer, J. J., and Krygier, J. J., "Fibrous Tissue Ingrowth and Attachment to Porous Tantalum," *J of Arthroplasty*, Vol. 14, No. 3, April 1999, pp. 347–354.

[5]   Medlin, D. J., Charlebois, S., Swarts, D., Shetty, R., and Poggie, R.A., "Metallurgical Characterization of a Porous Tantalum Biomaterial (Trabecular Metal™) for Orthopedic Implant Applications," *Proceedings from Materials and Processes for Medical Devices*, 8–10 September, 2003, ASM International, 2004, pp. 394–398.

[6]   *Annual Book of ASTM Standards*, Section Thirteen, "Medical Devices and Services," Vol. 13.01, ASTM International, West Conshohocken, PA, 2002.

[7]   *Annual Book of ASTM Standards*, Section Three, "Metals Test Methods and Analytical Procedures," Vol, 03.01, ASTM International, West Conshohocken, PA, 2003.

[8]   G. Vander Voort, *Metallography: Principles and Practice*, ASM International, 1984.

[9]   Medlin, D. J. and Compton, R., "Metallography of Biomedical Orthopedic Alloys," ASM Handbook, Vol. 9, *Metallography and Microstructures*, 10th ed., 2004.

[10] Samuels, L. E., *Metallographic Polishing by Mechanical Methods*, 3rd ed., ASM International, 1985.

[11] Medlin, D. J., Charlebois, S. J., Clarke, W., Pletcher, D. L., Scrafton, J. G., Shetty, R., et al., "Method for Attaching a Porous Metal Layer to a Metal Substrate," US Patent Application #20030232124, December 18, 2003.

[12] Donachie, M. J., "Titanium: A Technical Guide," ASM International, December 2000, pp. 13–24.

[13] Medlin, D. J., Vander Voort, G. F., and Lucas, G. M., "Metallographic Preparation of Orthopedic Medical Devices," *Proceedings from Materials and Processes for Medical Devices*, 25–26, August, 2004, ASM International.

[14] Metals Handbook, 8th ed., Vol. 8, "Metallography, Structures and Phase Diagrams," ASM International, p. 336.

[15] *Guidance Document for Testing Orthopedic Implants with Modified Metallic Surfaces Apposing Bone or Bone Cement*, US Food and Drug Administration, Office of Device Evaluation, Center for Devices and Radiological Health, April 28, 1994.

Journal of ASTM International, September 2005, Vol. 2, No. 8
Paper ID JAI12776
Available online at www.astm.org

Victor R. Jablokov,[1] *Naomi G. D. Murray,* [2] *Henry J. Rack,*[3] *and Howard L. Freese*[4]

Influence of Oxygen Content on the Mechanical Properties of Titanium-35Niobium-7Zirconium-5Tantalum Beta Titanium Alloy

ABSTRACT: Titanium-35Niobium-7Zirconium-5Tantalum beta titanium alloy (also known as TiOsteum® beta titanium alloy) is a new metallic biomaterial that was designed to have outstanding osseointegratabilty, good mechanical strength, improved ductility, very low elastic modulus, and good hot and cold workability [1]. In this study, mechanical property data are presented from several TiOsteum alloy production lots with ingot oxygen contents ranging from 0.05 to 0.68 %. Additionally, mill annealed mechanical property data collected over the last 10 to 15 years for the more common alpha, alpha-beta, and beta titanium alloys are presented as a function of oxygen content and compared to TiOsteum alloy. The influence of oxygen as an interstitial strengthening element is well established in titanium alloys. This study examines this effect over a broad range of oxygen values, and has significance for the application of the new TiOsteum alloy and other beta titanium alloys being considered for use in medical and surgical devices.

KEYWORDS: metals (for surgical implants), orthopedic medical devices, titanium alloys, beta titanium alloys, alpha-beta titanium alloys

Introduction

There are approximately 30 different metallic biomaterials that have been used or that are being considered for use to manufacture implantable medical and surgical devices [2]. These distinctly different metallic biomaterials are differentiated by their chemical compositions, and the resulting mechanical and metallurgical properties as defined by international ASTM Standards [3], ISO Standards [4], and UNS designations [5]. This list of metallic biomaterials is separated into four groups: stainless steels (iron-base alloys); cobalt-base alloys; titanium grades; and specialty grades.

Before the advent of implantable orthopedic and cardiovascular devices, metallic materials had been first developed for corrosion resistance and heat-resistance in other industries. Improved corrosion resistant stainless steels for the chemical industry, and cobalt-base alloys for the aerospace industry are examples of cross-industry application of metallurgical technology to the earliest medical implants for total joint arthroplasty. Dr. John Charnley's pioneering work with stainless steel hip stems in the 1960s was followed by experimentation with titanium and zirconium materials. Those early materials that were proven in successful device applications were defined in the first ASTM F04 "metallurgical materials" standards (ASTM subcommittee

Manuscript received 13 September 2004; accepted for publication 3 January 2005; published September 2005. Presented at ASTM Symposium on Titanium, Niobium, Zirconium, and Tantalum for Medical and Surgical Applications on 9-10 November 2004 in Washington, DC.
[1] Senior Engineer, ATI Allvac, an Allegheny Technologies company, Monroe, NC 28110.
[2] Research Engineer, Stryker Orthopaedics, Mahwah, NJ 07430.
[3] Professor of Materials Science and Engineering, Clemson University, Clemson, SC 29634.
[4] Business Development Manager, ATI Allvac, an Allegheny Technologies company, Monroe, NC 28110.

F04.12), and those standards were derived from published chemical industry and aerospace industry standards. These early "medical" materials were later designated as "grandfathered" material grades in ASTM F 763 (Table 1) and are considered, each on its own merit, as a reference material against which any new implantable metallic biomaterial must be compared [6].

TABLE 1—*Reference metallic biomaterials according to ASTM F 763.*

Common Name	ASTM & ISO Standards	UNS Number
Unalloyed Titanium; Grades CP-1, 2, 3, & 4	ASTM F 67, ISO 5832-2	R50250, 400, 550, & 700
Co-28Cr-6Mo Castings and Casting Alloy	ASTM F 75, ISO 5832-4	R30075
Co-20Cr-15W-10Ni-1.5Mn ("L-605") Alloy	ASTM F 90, ISO 5832-5	R30605
Ti-6Al-4V ELI Alloy	ASTM F 136, ISO 5832-3	R56401
Fe-18Cr-14Ni-2.5Mo ("316 LS") Alloy	ASTM F 138, ISO 5832-1	S31673
35Co-35Ni-20Cr-10Mo ("MP-35N") Alloy	ASTM F 562, ISO 5832-6	R30035

In the last 15 years, there have been important additions of new alloys to each of the four basic metals groups, as improved and totally new devices and applications have been developed. Three newer wrought stainless steel alloys are now being used in approved medical and surgical devices. Criteria for these new stainless steel grades were better corrosion fatigue properties, reduced nickel content, and similar or improved ductility. Two of these alloys have been patented, and those patents have since expired [7,8]. The latter alloy is noteworthy because of its significantly reduced nickel content, an important consideration for fracture fixation devices and for hypoallergenic jewelry.

- Fe-21Cr-12.5Ni-5Mn-2.5Mo ("XM-19," ASTM F 1314, S20910)
- Fe-22Cr-10Ni-3.5Mn-2.5Mo ("REX 734," ASTM F 1586, S31695)
- Fe-23Mn-21Cr-1Mo-1N ("108," ASTM F 2229, S29108)

Some important alloy development projects for cobalt-base alloy systems have resulted in novel chemistry and processing advances, and some improved cobalt-base alloys. First came the application of an older alloy that had been used as a spring wire in the Swiss watch industry, followed by two fairly similar grades [9,10]. Then, three variations on the cast Co-28Cr-6Mo alloy were made, and they are all now covered by a new wrought CoCrMo alloy standard, ASTM F 1537 (this standard was an outgrowth of the ASTM F 799 standard which was originally for a forging and machining alloy with chemistry almost identical to the ASTM F 75 standard which is for the casting alloy and castings). Alloy #3 in the ASTM F 1537 standard represents a CoCrMo grade with small additions of aluminum and lanthanum oxides. Patents for this Gas-Atomized-Dispersion-Strengthened (GADS) alloy relate to methods of manufacture and to improved properties of the alloy in the forged and sintered conditions [11]. More recently, several patents have been issued for "an alloy within an alloy;" a single-phase ASTM F 1537 Alloy #1 with improved high cycle fatigue properties [12]. Similarly, a higher fatigue version of the 35Co-35Ni-20Cr-10Mo (ASTM F 562) alloy has been introduced for wrought and drawn product forms, another "alloy within an alloy" with patents pending [13].

- Co-20Ni-20Cr-5Fe-3.5Mo-3.5W-2Ti ("Syncoben," ASTM F 563, R30563)
- Co-20Cr-15Ni-15Fe-7Mo-2Mn ("Elgiloy," ASTM F 1058, R30003)
- Co-19Cr-17Ni-14Fe-7Mo-1.5Mn ("Phynox," ASTM F 1058, R30008)

- Co-28Cr-6Mo ("GADS", ASTM F 1537, Alloy #3, R31539)
- Co-28Cr-6Mo ("No-Carb", ASTM F 1537, Alloy #1, R31537)
- 35Co-35Ni-20Cr-10Mo ("35N LT," ASTM F 562)

The greatest changes have occurred in the use of titanium and titanium alloys. The number of new titanium materials and product forms the medical device designer has from which to select. Since the early 1990s, six new ASTM standards for titanium-base alloys have been developed by the "Metallurgical Materials" Subcommittee, ASTM F-04.12, and these consensus standards have been balloted and approved by the "Medical and Surgical Materials and Devices" Main Committee, F-04 (Table 2). The first of these standards, ASTM F 1295, was an alpha + beta titanium alloy that was invented in Switzerland and was about to come "off patent" [14]. This new alpha + beta metallic biomaterial has intrinsic properties similar to the two "Ti-6-4" alloys but uses niobium instead of vanadium as a beta stabilizing alloying element. The second new standard, ASTM F 1472, was for the most widely produced aerospace titanium grade, Ti-6Al-4V alloy (UNS R56400).

ASTM F 1713 and F 1813, working through subcommittees simultaneously, were for two entirely new metastable beta titanium alloys with properties designed by medical device manufacturing companies specifically for structural orthopedic implant applications. Ti-13Nb-13Zr and Ti-12Mo-6Zr-2Fe are covered by one or more US and foreign patents [15,16]. The ASTM F 2066 standard was developed for the generic metastable beta titanium alloy, Ti-15Mo, and ASTM F 2146 covers the low-alloy alpha + beta Ti-3Al-2.5V tubing used for medical devices, which is based on the same product that has been used for aerospace hydraulic tubing for more than 40 years.

TABLE 2—*Titanium metallic biomaterials not referenced in ASTM F –763.*

Common Name	ASTM & ISO Standards	Microstructure	UNS Designation
Ti-5Al-2.5Fe Alloy ("Tikrutan)	ISO 5832-10	$\alpha + \beta$	Unassigned
Ti-6Al-7Nb Alloy ("TAN")	ASTM F 1295, ISO 5832-11	$\alpha + \beta$	R56700
Ti-6Al-4V Alloy	ASTM F 1472, ISO 5832-3	$\alpha + \beta$	R56400
Ti-13Nb-13Zr Alloy	ASTM F 1713	Metastable β	R58130
Ti-12Mo-6Zr-2Fe Alloy ("TMZF")	ASTM F 1813	Metastable β	R58120
Ti-15Mo Alloy	ASTM F 2066	Metastable β	R58150
Ti-3Al-2.5V Alloy (tubing only)	ASTM F 2146	$\alpha + \beta$	R56320
Ti-35Nb-7Zr-5Ta Alloy ("TiOsteum")	Sub. F-04.12.23	Metastable β	R58350

Another metastable beta titanium alloy, Ti-35Nb-7Zr-5Ta, was developed specifically for structural orthopedic implants such as total hip and total knee systems with the objectives of overcoming some of the technical limitations of the three established alpha + beta titanium alloys. With titanium, niobium, zirconium, and tantalum as alloying elements, the superior corrosion resistance and osseointegratabilty of this alloy were well demonstrated [17,18]. What has not been documented is the extent to which this quaternary alloy system can be engineered through alloy melting and semi-finished mill product processing techniques to provide a broad spectrum of properties for medical and surgical device applications. The objective of this paper is to consider the influence of oxygen as an interstitial strengthening element in titanium and titanium alloys, with special emphasis on the class of metastable beta titanium alloys and, in particular, the new highly-alloyed Ti-35Nb-7Zr-5Ta system.

Chemistry of Titanium-Base Metallic Biomaterials

Starting with the chemical requirements for the important titanium and titanium alloy grades, Table 3 was developed for side-by-side comparisons using the ASTM standards discussed above. For each grade, minima and maxima are listed for every specified alloying element, interstitial, and trace-level impurity element (if any). It is instructive to see the chemical compositions compared in this way and to observe that, in general, the higher maximum oxygen specification limits are associated with the grades with the greater the alloy contents. An artificial but meaningful measure of the alloy content is obtained by calculating the "Titanium, average" value that we define as the arithmetic average of the specified minimum and maximum limits of titanium content (by difference), according to the appropriate ASTM standard. Subtracting this value from unity, we obtain a measure of the alloy content (which includes interstitials).

These data demonstrate numerically the differences between the Ti CP titanium grades (alpha phase microstructure), the three alpha + beta titanium alloys, and three metastable beta titanium alloys. Although there are important chemical, mechanical, corrosion-resistance, and osseointegratabilty differences between the four CP titanium grades, we have chosen to represent this group as a single data point, Ti CP-4 (UNS R50700), so the differences in the other grades can be more easily seen.

Oxygen Content of Titanium-Base Metallic Biomaterials

Oxygen content influences the strength and ductility levels of the four CP titanium grades, with a doubling of oxygen from 0.18 % for Ti-CP-1 to 0.40 % for Ti-CP-4 driving an almost threefold increase in the specification minimum yield strength (from 172 MPa to 483 MPa). Elongation decreases from 24 % for Ti-CP-1 to 15 % for Ti-CP-4.

There are differences in both oxygen and alloy contents for the three alpha + beta titanium alloys: Ti-6Al-4V ELI and Ti-6Al-4V have maximum oxygen content and minimum specification yield strength values of 0.13 % and 795 MPa and 0.20 % and 860 MPa, respectively. Ti-6Al-7Nb is slightly more highly alloyed than Ti-6Al-4V and Ti-6Al-4VELI (13 % vs. 10 %), and it has an oxygen content of 0.20 % and a minimum specification yield strength requirement of 800 MPa.

Three metastable beta titanium alloys are a part of this analysis, two from the Ti-Mo group of alloys, and one Ti-Nb alloy. Both the oxygen and alloy content values are larger for these three alloys, and this is generally true for other commercially available metastable beta titanium alloys used in the aerospace industry; particularly so for Ti-3Al-8V-6Cr-4Mo-4Zr (UNS R58640) with maximum oxygen content and alloy content values of 0.25 % and 25 %, respectively. The three metastable beta titanium alloys for medical and surgical devices are: Ti-12Mo-6Zr-2Fe (UNS R58120), Ti-15Mo (UNS R58150), and Ti-35Nb-7Zr-5Ta (R58350). These newer alloys have alloy content values of 20 %, 15 %, and 47 % respectively. Table 4 summarizes the specified minimum and maximum oxygen levels for all three of these grades, along with values for the three alpha + beta alloys and Ti-CP-4. Note that the maximum oxygen content values for Ti-12Mo-6Zr-2Fe and Ti-35Nb-7Zr-5Ta are considerably greater than the three alpha + beta alloys. The maximum oxygen content value for Ti-15Mo, 0.20 %, may have been established at an inappropriately low level which is more consistent with the three alpha + beta alloys used for high fatigue strength medical device applications and for other alpha + beta alloys that are not used for medical and surgical devices.

TABLE 3—*Specification chemistry limits for commercially pure, alpha + beta and metastable beta titanium alloys.*

Element	Ti-CP-4 ASTM F 67[a] Gr. 4 (min)	(max)	Ti-6Al-4V ELI ASTM F 136 (min)	(max)	Ti-6Al-7Nb ASTM F 1295 (min)	(max)	Ti-6Al-4V ASTM F 1472 (min)	(max)	Ti-12Mo-6Zr-2Fe ASTM F 1813 (min)	(max)	Ti-15Mo ASTM F 2066 (min)	(max)	Ti-35Nb-7Zr-5Ta F 04.12.23 (min)	(max)
Nitrogen		0.05		0.05		0.05		0.05		0.05		0.05		0.02
Carbon		0.08		0.08		0.08		0.08		0.05		0.10		0.02
Hydrogen		0.015		0.012		0.009		0.015		0.020		0.015		0.020
Iron		0.50		0.25		0.25		0.30	1.50	2.50		0.10		0.25
Oxygen		0.40		0.13		0.20		0.20	0.008	0.28		0.20		0.75
Aluminum			5.50	6.50	5.50	6.50	5.50	6.75						
Vanadium			3.50	4.50			3.50	4.50						
Yttrium								0.005						
Niobium					6.50	7.50							34.00	37.00
Molybdenum									10.00	13.00	14.00	16.00		
Zirconium									5.00	7.00			6.30	8.30
Tantalum						0.50							4.50	6.50
Titanium[b] (by Δ)	100.00	98.96	91.00	88.478	88.00	84.91	91.00	88.10	83.49	77.10	86.00	83.54	55.20	47.14
Total	100.00	100.00	100.00	100.00	100.00	100.00	100.00	100.00	100.00	100.00	100.00	100.00	100.00	100.00
Titanium, ave.[c]	99.48 %		89.74 %		86.46 %		89.55 %		80.30 %		84.77 %		51.17 %	
Ave. Alloy Content	0.52 %		10.26 %		13.54 %		10.45 %		19.70 %		15.23 %		48.83 %	

[a] ASTM F 67 provides for four different compositions; Grade 1 ("CP-1"), Grade 2 ("CP-2"), Grade 3 ("CP-3"), and Grade 4 ("CP-4").
[b] Titanium content is determined by difference for values in both the minimum ("min") and the maximum ("max") columns.
[c] The "Titanium, ave." value is the arithmetic average of the theoretical minimum and maximum limits of titanium content according to the appropriate ASTM standard.

TABLE 4—*Specification average titanium and oxygen contents for eight titanium grades.*

Common Name	UNS Designation	Oxygen, min.	Oxygen, max.	Oxygen, ave.	Titanium, ave.
Ti CP-4	R50700	0.0	0.40	0.20	99.48
Ti-6Al-4V ELI	R56401	0.0	0.13	0.065	89.74
Ti-6Al-7Nb	R56700	0.0	0.20	0.10	86.46
Ti-6Al-4V	R56400	0.0	0.20	0.10	89.55
Ti-12Mo-6Zr-2Fe	R58120	0.008	0.28	0.144	80.30
Ti-15Mo	R58150	0.0	0.20	0.10	84.77
Ti-35Nb-7Zr-5Ta	R58350	0.0	0.75	0.375	51.17

Chemistry of Ti-35Nb-7Zr-5Ta

The Nb/Ta ratio versus Zr content of the Ti-35Nb-7Zr-5Ta alloy is represented in a two-dimensional diagram (Fig. 1) that assumes a four-component alloy system based on the alloy development work by the principal investigators [19]. This work performed on a multiplicity of button heats, as outlined in the patent documentation, has been supplemented by the melting of a number of production-scale heats. This supplemental work is characterized by the introduction of a fifth alloying element (oxygen) that is usually thought of as an important interstitial element that must be controlled in a fairly narrow range. Therefore, the intrinsic properties of an alloy with a practical chemical composition of about Ti-35Nb-7Zr-5Ta-(0.05-0.75)O are investigated. Data presented in this paper cover the oxygen range bounded by 0.05 to 0.68 %, a bit more narrow than the specification limits of 0.00 to 0.75 %.

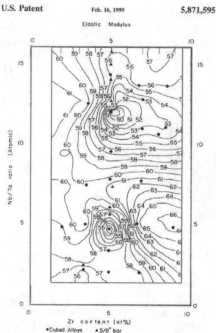

FIG. 1—*Modulus map for Ti-35Nb-7Zr-5Ta based on Nb/Ta ratio vs. Zr content.*

To achieve the optimum balance of mechanical properties, the melted-in chemistry of the production heats achieved the correct Nb/Ta ratio (atomic percent) and expected zirconium (atomic percent) so as to obtain a minimum elastic modulus (steep iso-contour lines of constant moduli in Fig. 1). The only significant difference from one heat to the next was the oxygen content and the fact that the finished bar products made from the different heats were processed in different production lots at different times. Therefore, properties of different lots of bar product are comparable with any significant differences being attributable to the different aims (and results) for the oxygen content.

Oxygen as an Interstitial Strengthener

The influence of oxygen as an interstitial strengthening element in titanium alloys is well known in alpha and alpha + beta titanium alloys. It is also known and accepted that the class of metastable beta titanium alloys has a generally higher affinity for interstitial elements such as oxygen, nitrogen, and hydrogen; and that the deleterious effects of nitrogen and hydrogen on mechanical properties is not as severe as for the CP titanium grades and the alpha + beta titanium alloys. Further, it is a fact that the beta titanium grades are generally much more highly alloyed than are the alpha + beta titanium grades. Lastly, it is well known that the metastable titanium alloys are generally more ductile as a class than are the alpha + beta titanium alloys.

Two metastable titanium alloys have been used successfully in aerospace for many years where excellent high-cycle fatigue strength, improved corrosion resistance, and superior ductility are demanded. These alloys are Ti-10V-2Fe-3Al [20] and Ti-3Al-8V-6Cr-4Mo-4Zr [21]. Ti-10V-2Fe-3Al is the preferred material for landing gear for commercial and military aircraft, and Ti-3Al-8V-6Cr-4Mo-4Zr is the preferred alloy for the large coil springs in landing gear for commercial and military aircraft. Since these jet aircraft components are exposed to the spray of saltwater on takeoff and landing, the military and aerospace standards require mechanical reliability in high cycle corrosion fatigue testing that is at least as severe as that required for highly loaded hip, knee, and fracture fixation implants *in vivo*. In addition, Ti-3Al-8V-6Cr-4Mo-4Zr alloy is used for high-performance racing (including NASCAR) and in some limited-production models of high-end European automobiles.

Correlating Yield Strength and Oxygen Content for Production Ingots

Most of the titanium semi-finished mill products delivered into the medical and surgical device pathways is manufactured in very large mill production lots as either round billet, round bar, round rod (small diameter bar cut to length), or rod coil stock for re-draw applications (such as wire and bone plate stocks). Similarly, most of the Ti-3Al-8V-6Cr-4Mo-4Zr alloy for aerospace and automotive applications is also manufactured as semi-finished long product by the titanium mills or their converters, whereas others produce finished goods from these so-called long products (as opposed to "flat products," which includes sheet, plate, and strip product forms). The Ti-10V-2Fe-3Al alloy is manufactured predominantly as a round "billet" product, a large diameter intermediate product that can be forged directly into the large truck beam components in landing gear assemblies. Some Ti-10V-2Fe-3Al alloy, however, is manufactured in the long product form and is used for brake rods in commercial aircraft.

The authors have been able to undertake an investigation using analytical data from production laboratory systems that have never been correlated in such a manner before. One manufacturer of titanium semi-finished mill product has production experience with each of the

CP, alpha + beta, and metastable beta titanium materials mentioned above for both aerospace and biomedical applications. The program team searched proprietary laboratory on-line files for the seven ASTM compositions listed in Table 4 for the same condition, processed on the same or similar equipment, generally using the same production routes. By sorting through a large body of data, a large sample would be obtained that would yield a statistically meaningful correlation of yield strength versus the important variable, oxygen content.

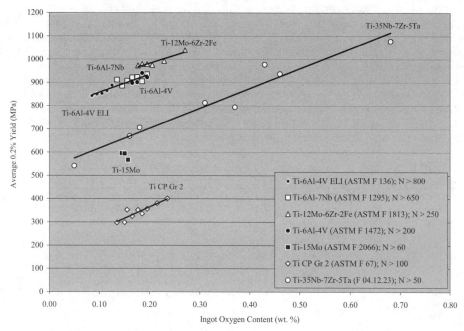

FIG. 2—*Dependence of average yield strength on the ingot oxygen content for titanium and titanium alloys (alpha, alpha + beta, and metastable beta microstructures).*

The influence of ingot oxygen content on the *average* yield strength of various titanium/titanium alloy metallic biomaterials is shown in Fig. 2. Each data point represents a "batch" of consolidated and averaged yield data from one or numerous ingots/heats having identical ingot oxygen content. The ingot oxygen content listed for each data point is the certified ingot oxygen level. Figure 2 reveals a comparison of mill product data in the *mill annealed* condition for various round bar product diameters that, as mentioned above, have been similarly manufactured (PAM or VAR melted, press and rotary forged to intermediate billet, hot rolled to round bar/coil and finish machined [22]) and conform to the applicable biomedical specifications. The corresponding *average* data are listed in Table 5, and the standard error computed by regression analysis (a measure of the data spread) is listed in Table 6.

The comparison shown in Fig. 2 is meant to be a "macro" representation of the influence of oxygen content on the yield properties of various titanium/titanium alloys. Therefore, as mentioned above, each data point represents the average of all yield strength data collected for each oxygen content and ignores minor variances in processing parameters such as rolling temperature, mill anneal temperature, final bar size, etc. Subsequently, over 2000 data sets have been analyzed to generate Fig. 2. Not surprisingly, the figure confirms the recognized dependence of yield strength on the interstitial strengthener, oxygen, for titanium and titanium alloys [23,24]. Namely, as the oxygen level increases so does the yield strength. Additionally, Fig. 2 allows the interstitial strengthening contribution of oxygen to be predicted over a range of ingot oxygen levels for various titanium alloys.

Comparing the Ti-35Nb-7Zr-5Ta metastable beta alloy in the mill annealed condition to the other alpha + beta, commercially pure, and metastable beta alloys shown in Fig. 2 the following is observed. For oxygen levels between 0.16 % to ~0.38 %, Ti-35Nb-7Zr-5Ta has lower yield strength than all of the alloys depicted other than Ti CP Gr. 2 (commercially pure) and Ti-15Mo metastable beta alloy. For oxygen levels between 0.38–0.62 %, the range in yield strength for Ti-35Nb-7Zr-5Ta is equivalent to the yield strength range of the alpha + beta (Ti-6Al-4V ELI, Ti-6Al-4V and Ti-6Al-7Nb) and Ti-12Mo-6Zr-2Fe metastable beta alloy in the figure. For oxygen levels above ~0.62 %, the yield strength of Ti-35Nb-7Zr-5Ta exceeds that of all of the other alloys. As a result, a broad range of yield strength is achievable for Ti-35Nb-7Zr-5Ta simply by increasing or decreasing the ingot oxygen content.

TABLE 5—*Average yield strength data as a function of ingot oxygen content for commercially pure, alpha + beta and metastable beta titanium alloys.*

Ti-35Nb-7Zr-5Ta		Ti-15Mo		Ti-12Mo-6Zr-2Fe		Ti CP Gr 2	
Ingot O (wt. %)	Ave. Yield (MPa)	Ingot O (wt. %)	Ave. Yield (MPa)	Ingot O (wt. %)	Ave. Yield (MPa)	Ingot O (wt. %)	Ave. Yield (MPa))
0.05	542	0.14	596	0.18	972	0.14	297
0.16	669	0.15	594	0.19	979	0.15	299
0.18	706	0.16	568	0.20	978	0.16	353
0.31	813			0.21	974	0.17	325
0.37	794			0.23	992	0.18	352
0.43	977			0.27	1038	0.19	336
0.46	937					0.20	356
0.68	1078					0.22	381
St. Err.	30					0.24	401

Ti-6Al-7Nb		Ti-6Al-4V		Ti-6Al-4V ELI	
Ingot O (wt. %)	Ave. 0.2 % Yield (MPa)	Ingot O (wt. %)	Ave. 0.2 % Yield (MPa)	Ingot O (wt. %)	Ave. 0.2 % Yield (MPa)
0.14	911	0.17	897	0.09	843
0.15	886	0.18	901	0.10	850
0.16	907	0.19	940	0.11	853
0.17	921	0.20	921	0.12	864
0.18	922			0.13	887
0.19	904				
0.20	934				

TABLE 6—*Standard error computed by performing regression analysis of the raw data for each alloy system shown in Fig. 2.*

	Ti-35Nb-7Zr-5Ta	Ti-15Mo	Ti-12Mo-6Zr-2Fe	Ti CP Gr 2	Ti-6Al-7Nb	Ti-6Al-4V	Ti-6Al-4V ELI
St. Err. (MPa)	±30	N/a	±22	±29	←	±42	→

Mechanical Properties of Mill Annealed Ti-35Nb-7Zr-5Ta

A more detailed view of Ti-35Nb-7Zr-5Ta tensile data is shown in Fig. 3. In this figure, ultimate tensile stress, yield stress, elongation, and reduction of area are shown as a function of ingot oxygen content. As in Fig. 2, each data column/point consists of an average of all available mill annealed test data from various mill product forms for a specific ingot oxygen level.

In Fig. 3, again the dependence of strength on oxygen content is reiterated as was shown in Fig. 2. As the oxygen level increases from 0.16–0.68 %, the ultimate tensile strength of the alloy increases from ~715 MPa to ~1096 MPa, and the yield strength increases from ~669 MPa to ~1077 MPa (see Table 7). Additionally, it is interesting to note that ductility does not decrease as the strength increases with increasing ingot oxygen content, a finding that has been reported in other work as well [25]. The ductility of Ti-35Nb-7Zr-5Ta is greater than 18.5 % throughout the entire oxygen range that was studied.

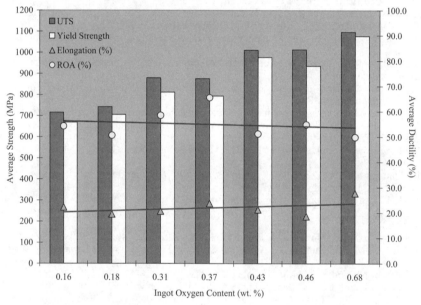

FIG. 3—*Ti-35Nb-7Zr-5Ta tensile properties versus ingot oxygen content.*

TABLE 7—*Ti-35Nb-7Zr-5Ta tensile properties versus ingot oxygen content.*

Ingot Oxygen (wt. %)	Yield (MPa)	UTS (MPa)	Elongation (%)	ROA %)
0.16	669	715	22.2	54.3
0.18	706	742	19.5	50.6
0.31	812	880	20.7	58.5
0.37	876	794	23.7	65.5
0.43	977	1011	21.3	51.2
0.46	936	1013	18.7	54.8
0.68	1077	1096	27.7	49.9

Conclusions

- Yield strength and ultimate tensile strength of Ti-35Nb-7Zr-5Ta and other titanium alloys increases with increasing ingot oxygen content.
- A broad range of yield and ultimate tensile strengths spanning ~660 to ~1096 MPa can be produced for Ti-35Nb-7Zr-5Ta simply by increasing or decreasing the ingot oxygen content.
- The ductility of Ti-35Nb-7Zr-5Ta is not negatively affected if strength is increased due to increasing oxygen content.

References

[1] Ahmed, T. and Rack, H. J., "Low Modulus Biocompatible Titanium Base Alloys for Medical Devices." US Patent Number 5,871,595 (Feb. 16, 1999); Ahmed, T. and Rack, H. J., "Low Modulus, Biocompatible Titanium Base Alloys for Medical Devices," European Patent Specification EP 0 707 085 B1 (07.01.1999).

[2] Lemons, J. and Freese, H., "Metallic Biomaterials For Surgical Implant Devices," *BoneZone*, (a Knowledge Enterprises, Inc. publication), Fall 2002, pp. 5–9.

[3] *Annual Book of ASTM Standards*, Vol. 13.01, ASTM International, West Conshohocken PA.

[4] International Organization for Standardization, ISO Central Secretariat, Geneva, Switzerland.

[5] *Metals and Alloys in the Unified Numbering System (UNS)*, 9th ed., ASTM International, West Conshohocken, PA.

[6] ASTM F 763, "Standard Practice for Short-Term Screening of Implant Materials," *Annual Book of ASTM Standards*, ASTM International, West Conshohocken, PA.

[7] BS 7252: Part 9: 1990, "Metallic Materials for Surgical Implants: Part 9, Specification for High-Nitrogen Stainless Steel," British Standards Institution (British Patent Number 1 154 934).

[8] US Patents Number 3 820 980, Number 3 847 599, Number 3 907 551, Number 3 936 297, Number 4 21137, Number 4 217 150.

[9] ASTM F 563, "Standard Specification for Wrought Cobalt-20Nickel-20Chromium-3.5Molybdenum-3.5Tungsten-5Iron Alloy for Surgical Implant Applications (UNS R30563)."

[10] ASTM F 1058, Standard Specification for Wrought 40Cobalt-20Chromium-16Iron-15Nickel-7Molybdenum Alloy Wire and Strip for Surgical Implant, *Annual Book of ASTM Standards*.

[11] US Patent Number 4 714 468, Number 4 687 290.

[12] US Patent Number 6 187 045, Number 6 539 607, Number 6 773 520, and European Patent Number WO 00/47140.

[13] Bradley, D., Kay, L., Stephenson, T. and Lippard, H., "Optimization of Melt Chemistry and   Properties of 35Cobalt-35Nickel-20Chromium-10Molybdenum Alloy (ASTM F 562) Medical Grade Wire," ASM International M&PMD Conference, Anaheim CA, Sept. 2003.

[14] ASTM F 1295, "Standard Specification for Wrought Titanium-6 Aluminum-7 Niobium Alloy for Surgical Implant Applications (UNS R56700)."

[15] ASTM F 1713, "Standard Specification for Wrought Titanium-13Niobium-13Zirconium Alloy for Surgical Implant Applications)," *Annual Book of ASTM Standards*.

[16] ASTM F 1813, "Standard Specification for Wrought Titanium-12 Molybdenum-6 Zirconium-2 Iron Alloy for Surgical Implant (UNS R58120)," current edition approved October 10, 2001.

[17] Hawkins, M. J., Ricci, J. L., Kauffman, J., and Jaffe, W., "Osseointegration of a New Beta Titanium Alloy as Compared to Standard Orthopaedic Implant Materials," No. 1083, Sixth

World Biomaterials Congress, Society for Biomaterials, May 2000.
[18] Shortkroff, S., Zhang, X. Y., Rice, K., Dimaano, F., Thornhill, T. S., "In Vitro Biocompatibility of TiOsteum," No. 341, Society for Biomaterials, Brigham and Women's Hospital and Harvard Medical School, April 2002.
[19] Ibid., US Patent Number 5 871 595.
[20] ASM 4984, "Titanium Alloy Forgings 10V-2Fe-3Al," Society of Automotive Engineers (SAE), Warrendale PA.
[21] ASM 4957, "Titanium Alloy, Round Bar and Wire 3Al – 8V – 6Cr – 4Mo – 4Zr, Consumable Electrode Melted, Solution Treated and Cold Drawn," Society of Automotive Engineers (SAE), 400 Commonwealth Drive, Warrendale PA.
[22] Davis, R. M. and Forbes-Jones, R. M., "Manufacturing Process for Semi-Finished Titanium Biomedical Alloys," *Medical Applications of Titanium and Its Alloys: The Material and Biological Issues,* S. A. Brown and J. E. Lemons, Ed., ASTM International, West Conshohocken, PA, 1996, p. 17.
[23] Imam, M. A. and Feng, C. R., "Role of Oxygen on Transformation Kinetics of Timetal-21S Titanium Alloy," *Advances in the Science and Technology of Titanium Alloy Processing,* TMS, I. Weiss, R. Srinivasan, P. J. Bania, D. Eylon, and S. L. Semiatin, Ed., Warrendale, PA, 1996, pp. 435–450.
[24] Qazi, J. I., Tsakiris, V., Marquardt, B. and Rack, H. J., "The Effect of Duplex Aging on the Tensile Behavior of Ti-35Nb-7Zr-5Ta-(0.06-0.7)O Alloys," *Ti-2003 Science and Technology, Proceedings of the 10th World Conference on Titanium,* WILEY-VCH Verlag GmbH & Co. KGaA, G. Lutjering and J. Albrecht, Ed., Weinheim, Germany, 2004, pp. 1651.
[25] Qazi, J. I., Tsakiris, V., Marquardt, B. and Rack, H. J., "Effect of Aging Treatments on the Tensile Properties of Ti-35Nb-7Zr-5Ta-(0.06-0.7)O Alloys," *Titanium, Niobium, Zirconium, and Tantalum for Medical and Surgical Applications, 2004, Nov 2004 Conference Proceedings* (in press), L. Zardiackas.

Journal of ASTM International, September 2005, Vol. 2, No. 8
Paper ID JAI12780
Available online at www.astm.org

J. I. Qazi,[1] *V. Tsakiris,*[1] *B. Marquardt,*[2] *and H. J. Rack*[1]

Effect of Aging Treatments on the Tensile Properties of Ti-35Nb-7Zr-5Ta-(0.06-0.7)O Alloys

ABSTRACT: Ti-35Nb-7Zr-5Ta rods containing 0.06, 0.46, and 0.68 wt.% oxygen were produced by vacuum arc melting and forging. These were solution treated (ST) and aged in the temperature range of 427–593°C for 8 h followed by air cooling. Increasing oxygen content increased the ST yield strength (YS) from 530 to 1081 MPa. The largest increase in YS, 630MPa, in Ti-35Nb-7Zr-5Ta-0.06O observed after aging at 427°C was associated with ω phase precipitation. While the largest increases in YS, (1060 and 1288 MPa, in Ti-35Nb-7Zr-5Ta-0.46/0.68O) observed after aging at 482°C were associated with fine and somewhat inhomogeneous precipitation of the α phase, the tensile elongation remained above 8 %. Aging at 538 or 593°C led to an increase in the inhomogeneity in α phase precipitation in Ti-35Nb-7Zr-5Ta-0.46/0.68O, precipitate free zones clearly being observed along with prior β grain boundaries, which served as preferred paths for early crack propagation and decreased tensile ductility.

KEYWORDS: biomaterials, titanium alloys, aging, mechanical properties, fractography

Introduction

Excellent biocompatibility, corrosion resistance, and low density make titanium and its alloys attractive materials for biomedical and dental applications [1–6]. Titanium alloys used for this purpose can be divided broadly into three categories: commercial purity α, α+β, and low modulus metastable β alloys, the latter having been developed to promote load sharing between the implant and natural bone. The tensile properties of commercial purity α-Ti alloys are primarily controlled by their interstitial content, grain size, grain morphology, and crystallography texture [7,8]. In contrast, the tensile properties of α+β Ti alloys can be tailored over a wider range by additionally controlling the size, morphology, and volume fraction of the α and β phases [7,9–12].

Finally, the low modulus metastable β alloys are generally used in the low strength solution treated condition. The yield strength of metastable β alloys, however, can be increased by increasing oxygen content and/or by aging [13–19]. This improvement makes these alloys excellent candidates for other biomedical applications, e.g., bone plates. The current study is part of a broader effort to understand the effect of oxygen and/or aging on the tensile properties of Ti-35Nb-7Zr-5Ta. A previous study [20] has shown for example that duplex aging (260°C 4 h/ 427°C 8 h) of Ti-35Nb-7Zr-5Ta-0.46O can be used to achieve strength levels as high as 1200 MPa with 8 % elongation. This investigation has examined the combined effects of oxygen content and aging in the temperature range of 427–593°C on the tensile properties of these

Manuscript received 13 August 2004; accepted for publication 26 January 2005; published September 2005.
Presented at ASTM Symposium on Titanium, Niobium, Zirconium, and Tantalum for Medical and Surgical Applications on 9-10 November 2004 in Washington, DC.
[1] School of Materials Science and Engineering, Clemson University, Clemson, SC 29634, USA.
[2] Formerly Allvac, Monroe, NC 28111, now Zimmer Inc., Warsaw, IN 46581, USA.

alloys. The precipitation phenomena associated with these changes in tensile properties are described in detail elsewhere [21].

Experimental

Ti-35Nb-7Zr-5Ta ingots with three different oxygen contents, 0.06, 0.46, and 0.68 wt.%, were produced by vacuum arc melting, the differing oxygen contents being controlled through addition of rutile (TiO$_2$) (Table 1). These ingots were hot forged, rolled to 16 mm diameter rods, and solution treated at 850°C (0.06 wt.% O), 840°C (0.46 wt.% O), and 900°C (0.68 wt.% O) for one hour followed by water quenching. Tensile properties per ASTM E8 at a strain rate of 0.5s^{-1} were then determined in the solution treated condition or after additional aging at 427, 482, 538, or 593°C for 8 h followed by air-cooling.

TABLE 1—*Chemical composition of Ti-35Nb-7Zr-5Ta alloys (wt.%).*

Alloy	Ti	Nb	Zr	Ta	H	C	N	O
Low O	Bal.	35.3	7.2	4.9	0.001	0.013	0.002	0.06
Medium O	Bal.	34.6	7.3	5.6	0.005	0.046	0.009	0.46
High O	Bal.	34.6	7.1	5.6	0.006	0.048	0.012	0.68

Microstructure analysis was performed using optical and scanning electron microscopy. Samples for this purpose were prepared by standard mechanical polishing with final etching in a solution consisting of 8 vol.% HF, 15 vol.% HNO$_3$, and 77 vol.% distilled H$_2$O. Phase identification was also undertaken by x-ray diffraction using Cu-Kα radiation at 45 kV and 40 mA, with a scanning rate of 0.06°/min over the range of 30–90°, the diffraction data being collected with a Peltier solid-state detector.

Finally, scanning electron microscopy of the failed tensile fracture surfaces was utilized to provide a qualitative description of the influence of increased oxygen content and aging upon the observed tensile failure phenomena.

Results

Solution Treated Condition

As described previously, all of the solution treated alloys examined in this investigation appeared to consist of a single β phase structure with an average grain size of 60, 23, and 28 μm respectively [20]. Figure 1 illustrates this microstructure condition for Ti-35Nb-7Zr-5Ta-0.06O, the black spots seen being common to all and representing etch pits that develop during the rather lengthy etching time required. X-ray analysis confirmed their single phase β-structure (Fig. 2) [20]. TEM analysis supported this observation, diffused scattering (an indication of early stage of ω formation) only being observed in the selected area diffraction patterns of Ti-35Nb-7Zr-5Ta-0.06O.

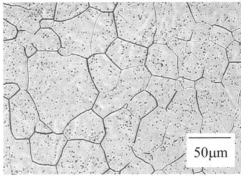

FIG. 1—*Optical photomicrograph of Ti-35Nb-7Zr-5Ta-0.06O in the solution treated*
condition.

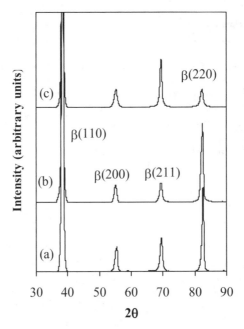

FIG. 2—*X-ray diffraction patterns of solution treated Ti-35Nb-7Zr-5Ta containing (a) 0.06,*
(b) 0.46, and (c) 0.68 wt.% oxygen.

The tensile properties of the Ti-35Nb-7Zr-5Ta-O alloys in the solution treated as well as in
the solution treated and aged conditions are summarized in Tables 2–4. In general, the yield and
ultimate tensile strengths of these alloys in the solution treated condition increased with
increasing oxygen content, the tensile ductility remaining essential unaffected by this increase.
Similarly, scanning electron microscopy showed that failure in the solution treated condition,
independent of oxygen content, involved ductile dimple transgranular fracture (Fig. 3).

TABLE 2—*Tensile properties of Ti-35Nb-7Zr-5Ta-0.06O.*

Heat Treatment	YS (MPa)	UTS (MPa)	El. (%)	RA (%)
Solution treated	530	590	21	69
427°C/ 8 h	630	686	17	42
482°C/ 8 h	503	534	20	46
538°C/ 8 h	493	537	21	52
593°C/ 8 h	508	549	26	63

TABLE 3—*Tensile properties of Ti-35Nb-7Zr-5Ta-0.46O.*

Heat Treatment	YS (MPa)	UTS (MPa)	El. (%)	RA (%)
Solution treated	937	1014	19	55
427°C/ 8 h	1007	1055	12	27
482°C/ 8 h	1060	1149	9	17
538°C/ 8 h	806	929	11	22
593°C/ 8 h	765	861	15	22

TABLE 4—*Tensile properties of Ti-35Nb-7Zr-5Ta-0.68O.*

Heat Treatment	YS (MPa)	UTS (MPa)	El. (%)	RA (%)
Solution treated	1081	1097	21	50
427°C/ 8 h	1222	1252	9	13
482°C/ 8 h	1288	1362	8	9
538°C/ 8 h	1036	1180	9	11
593°C/ 8 h	893	1036	10	12

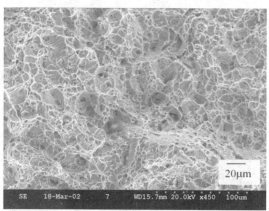

FIG. 3—*Scanning electron micrograph illustrating the tensile fracture surface of solution treated Ti-35Nb-7Zr-5Ta-0.68O.*

Aging at 427°C for 8 h

Analysis of Ti-35Nb-7Zr-5Ta-0.06O indicated that aging at 427°C for 8 h had little effect on the microstructure when compared to the solution treated condition (Fig. 4a). X-ray diffraction did reveal, however, the presence of the ω phase (Fig. 5a). Microstructural examination of Ti-35Nb-7Zr-5Ta–0.46/0.68O revealed the presence of fine α precipitates (Figs. 4b and c), its presence after aging at 427°C being confirmed by x-ray diffraction, as evidenced by the presence of (102), (201), and ($2\overline{1}2$) α phase peaks (Figs. 5*b* and *c*). X-ray diffraction also revealed the presence of ($2\overline{1}1$), ($2\overline{1}2$), and (301) ω phase peaks in Ti-35Nb-7Zr-5Ta-0.46O aged at 427°C for 8 h (Fig. 5*b*).

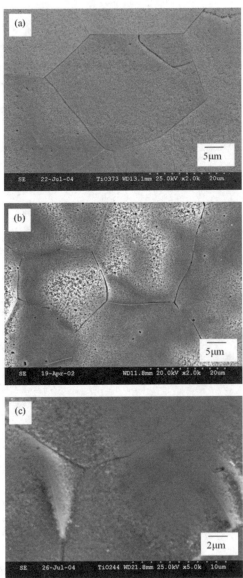

FIG. 4—*Scanning electron micrographs illustrating the microstructure of Ti-35Nb-7Zr-5Ta containing* (a) *0.06,* (b) *0.46, and* (c) *0.68 wt.% oxygen aged at 427 °C for 8 h.*

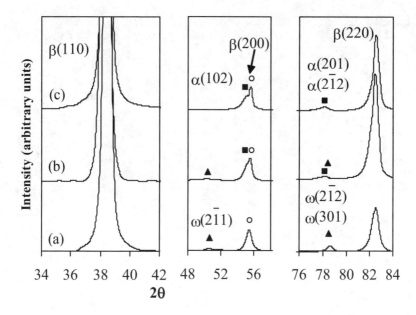

FIG. 5—*X-ray diffraction patterns of Ti-35Nb-7Zr-5Ta containing (a) 0.06, (b) 0.46, and (c) 0.68 wt.% oxygen, aged at 427°C for 8 h.*

The tensile results indicate that aging at 427°C for 8 h increases the yield strength of all the alloys examined when compared to the solution treated condition (Fig. 6). The largest percentage increase in yield strength, approximately 19 %, occurred in Ti-35Nb-7Zr-5Ta-0.06O alloy (Table 2). In contrast, aging of Ti-35Nb-7Zr-0.46O and Ti-35Nb-7Zr-5Ta-0.68O for 8 h at 427°C resulted in an increase of approximately 7.5 and 13 % in their respective yield strength vis a vis the solution treated condition (Tables 3 and 4). These increases in strength were accompanied by a decrease in tensile ductility, both the elongation and reduction in area decreasing for all three alloys after aging at 427°C for 8 h (Fig. 6). The percentage decrease in tensile ductility was dependent on the oxygen content, the largest percentage decrease being observed in the highest O alloy.

Scanning electron microscopy of the Ti-35Nb-7Zr-5Ta- 0.06O and 0.46O alloys' tensile fracture surfaces showed a mixture of ductile dimple (A) and smooth (B) areas (Figs. 7a and b), with a more detailed examination of the latter showing evidence of fine striations (C). Evidence of secondary cracking, presumably along prior beta grain boundaries, was also observed. Similar examination of the Ti-35Nb-7Zr-5Ta-0.68O tensile fracture surfaces showed a mixture of ductile dimple (A) and scalped or layered (D) areas (Fig. 7c).

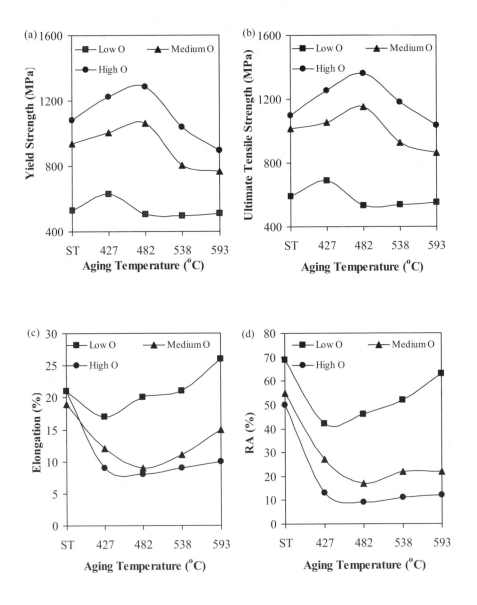

FIG. 6—*Effect of aging temperature on* (a) *yield strength,* (b) *ultimate tensile strength,* (c) *elongation, and* (d) *reduction in area for Ti-35Nb-7Zr-5Ta alloys.*

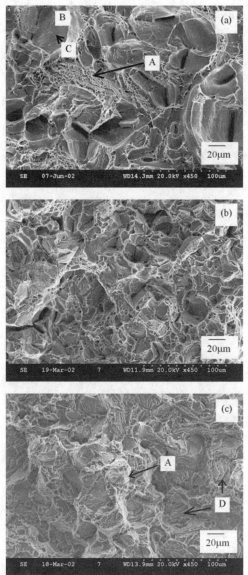

FIG. 7—*Scanning electron micrographs illustrating the tensile fracture surfaces of Ti-35Nb-7Zr-5Ta containing (a) 0.06, (b) 0.46, and (c) 0.68 wt.% oxygen aged at 427 °C for 8 h.*

Aging at 482°C for 8 h

Analysis of Ti-35Nb-7Zr-5Ta-0.06O again suggested that aging of this alloy at 482°C for 8 h had little effect on its microstructure (Fig. 8a). Indeed, x-ray diffraction also failed to reveal the presence of the ω phase after this aging treatment (Fig. 9a), that is the material was essentially a single phase β-Ti alloy. Examination of Ti-35Nb-7Zr-5Ta-0.46/0.68 O showed that the aging of these systems at 482°C for 8 h resulted in inhomogenous precipitation of fine α phase (Figs. 8b and c). Indeed, precipitate free zones, devoid of α phase, were observed to be associated with many grain boundaries, x-ray diffraction confirming the presence of α in these O alloys (Fig. 9).

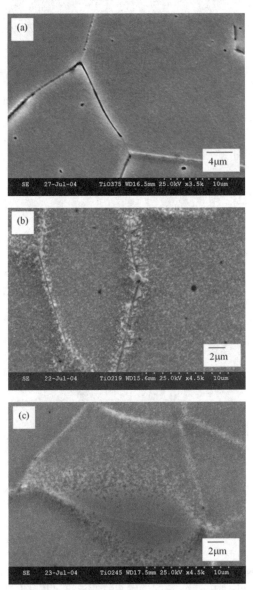

FIG. 8—*Scanning electron micrographs illustrating the microstructure of Ti-35Nb-7Zr-5Ta containing (a) 0.06, (b) 0.46, and (c) 0.68 wt.% oxygen aged at 482°C for 8 h.*

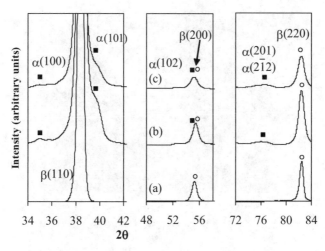

FIG. 9—*X-ray diffraction patterns of Ti-35Nb-7Zr-5Ta containing (a) 0.06, (b) 0.46, and (c) 0.68 wt.% oxygen aged at 482 °C for 8 h.*

The tensile results indicated that aging at 482°C for 8 h actually resulted in a slight decrease in the strength of Ti-35Nb-7Zr-5Ta-0.06O, without any appreciable change in tensile ductility. Aging of Ti-35Nb-7Zr-5Ta-0.46/0.68O under the same conditions resulted, respectively, in approximate 13 and 20 % increase in yield strength when compared to the solution treated material. Indeed, aging at 482°C resulted in the highest yield strengths observed with the tensile ductility still remaining within acceptable limits, i.e., 8 % elongation.

Scanning electron microscopy showed that the tensile fracture surface of Ti-35Nb-7Zr-5Ta-0.06O consisted of ductile dimple (A) failure and quasi-cleavage (E) regions (Fig. 10a). Similar observations made for Ti-35Nb-7Zr-5Ta-0.46/0.68O aged at 482°C indicated crack propagation involving a mixture of ductile dimple, quasi-cleavage and secondary cracking (Figs. 10b and c).

Aging at 538°C for 8 h

Analysis indicated that aging of Ti-35Nb-7Zr-5Ta-0.06O at 538°C for 8 h again had little effect on microstructure (Fig. 11a) single β-phase structure being confirmed by x-ray diffraction analysis (Fig. 12a). Similarly, aging of Ti-35Nb-7Zr-5Ta-0.46/0.68O under these conditions resulted in coarsening of the lenticular α phase precipitation previously observed at lower aging temperatures within the former β grains (Figs. 11b and c). Additionally, the α was again inhomogeneously distributed, precipitate free zones again being observed, the width of these zones appearing to increase with increasing aging temperature, x-ray diffraction confirming the presence of α+β phases in these alloys (Fig. 12).

The tensile results indicated that aging at 538°C for 8 h resulted in a decrease in the yield strength of all the Ti-35Nb-7Zr-5Ta alloys examined in this investigation, with the yield strength actually decreasing below that observed in the solution treated condition. However, some recovery in the tensile ductility with respect to aging at 482°C for 8 h was noted.

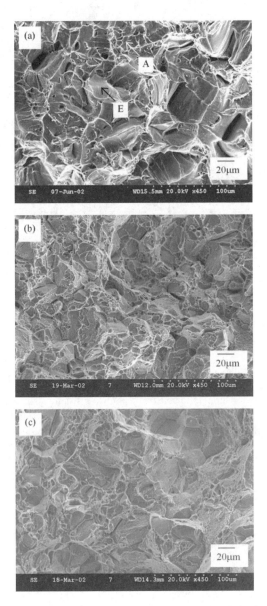

FIG. 10—*Scanning electron micrographs illustrating the tensile fracture surfaces of Ti-35Nb-7Zr-5Ta containing (a) 0.06, (b) 0.46, and (c) 0.68 wt.% oxygen aged at 482 °C for 8 h.*

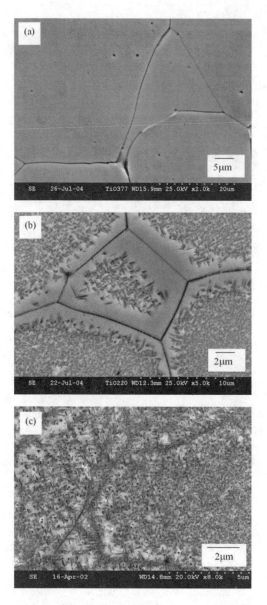

FIG. 11—*Scanning electron micrographs illustrating the microstructure of Ti-35Nb-7Zr-5Ta containing (a) 0.06, (b) 0.46, and (c) 0.68 wt.% oxygen aged at 538 °C for 8 h.*

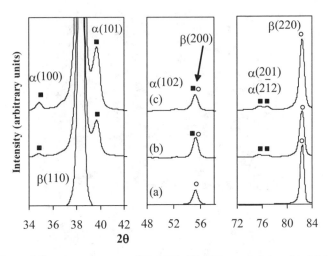

FIG. 12—*X-ray diffraction patterns of Ti-35Nb-7Zr-5Ta containing (a) 0.06, (b) 0.46, and (c) 0.68 wt.% oxygen aged at 538 °C for 8 h.*

Scanning electron microscopy showed that fracture of Ti-35Nb-7Zr-5Ta-0.06O aged at 538°C was associated with ductile dimple failure (Fig. 13a), while that observed in Ti-35Nb-7Zr-5Ta-0.46/0.68O consisted primarily of quasi-cleavage and fine ductile dimple and secondary cracking (Figs. 13b and c).

Aging at 593°C for 8 h

Once again, analysis indicated that aging of Ti-35Nb-7Zr-5Ta-0.06O at 593°C for 8 h had little effect on its microstructure (Fig. 14a), the alloy being single phase β as confirmed by x-ray diffraction analysis (Fig. 15a). Aging of the Ti-35Nb-7Zr-5Ta-0.46/0.68O under these same conditions resulted in a continued coarsening of the lenticular α phase precipitates observed within the prior β grains at lower temperatures, an increase in the precipitation free zone width, and the initiation of grain boundary α precipitation (Figs. 14b and c). The presence of α+β phases in these alloys was confirmed by x-ray diffraction analysis (Figs. 15b and c).

The tensile results indicated that increasing the aging temperature to 593°C had little effect on the properties of Ti-35Nb-7Zr-5Ta-0.06O, while it resulted in a continued decrease in the yield strength of Ti-35Nb-7Zr-5Ta-0.46/0.68O alloys. In all instances, these changes were again accompanied by an increase in the tensile elongation in comparison to aging at 538°C for 8 h. In fact, the tensile elongation of Ti-35Nb-7Zr-5Ta-0.06O aged at 593°C exceeded that observed in the solution treated condition. Scanning electron microcopy of tensile fracture surface of Ti-35Nb-7Zr-5Ta-0.06O showed a mixture of fine and coarse ductile dimples (Fig. 16a), while the tensile fracture surfaces of the Ti-35Nb-7Zr-5Ta-0.46/0.68O samples consisted of a mixture of quasi-cleavage and extremely fine ductile dimple areas, with secondary cracking running parallel to the prior tensile axis and associated flat smooth areas (Figs. 16b and c).

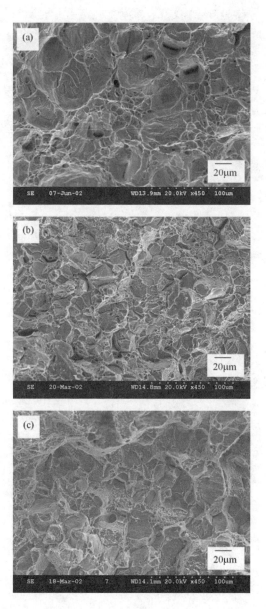

FIG. 13—*Scanning electron micrographs illustrating the tensile fracture surfaces of Ti-35Nb-7Zr-5Ta containing (a) 0.06, (b) 0.46, and (c) 0.68 wt.% oxygen aged at 538°C for 8 h.*

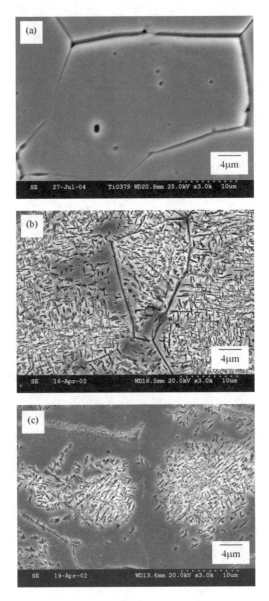

FIG. 14—*Scanning electron micrographs illustrating the microstructure of Ti-35Nb-7Zr-5Ta containing (a) 0.06, (b) 0.46, and (c) 0.68 wt.% oxygen aged at 593 °C for 8 h.*

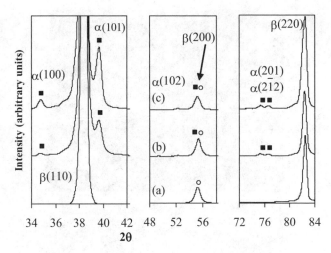

FIG. 15—*X-ray diffraction patterns of Ti-35Nb-7Zr-5Ta containing (a) 0.06, (b) 0.46, and (c) 0.68 wt.% oxygen aged at 593 °C for 8 h.*

Discussion

Aging at 427°C for 8 h

Aging at 427°C for 8 h resulted in an increase in yield and ultimate tensile strength independent of oxygen content. Microstructure and phase analysis showed that this increase in strength resulted from the precipitation of ω phase in Ti-35Nb-7Zr-5Ta-0.06O, a mixture of ω+α phases in Ti-35Nb-7Zr-5Ta-0.46O, and α phase in Ti-35Nb-7Zr-5Ta-0.68O. This observation suggests that increasing oxygen content tends to suppress ω phase formation, while promoting the formation of fine α and is discussed in detail in Ref. [21]. The conclusion that increasing oxygen tends to suppress ω phase formation while supporting α phase formation has also been noted by Williams et al. [22] in Ti-25V. In the latter study the authors reported that when Ti-25V was aged at temperatures between 300 and 450 °C, increasing the oxygen to 1 wt.% reduced the volume fraction of ω phase and reduced the time required for α phase formation.

The observed increases in the yield strength were accompanied with decreases in tensile ductility (Fig. 6). Fractographic results suggest that the observed decrease in Ti-35Nb-7Zr-5Ta-0.06/0.46O is associated with the formation of smooth fracture surface regions containing what appear to be slip lines or striations. Again, the formation of these may be attributed to the presence of the fine ω phase precipitates, which are know to promote planar slip and thereby limit ductility [23]. Indeed, further examination of Fig. 7 suggests that the intersection of these slip bands with favorably oriented grain boundaries tends to promote crack formation and crack propagation along prior β grain boundaries. The present study has also shown that when ω is no longer present, for example following aging of Ti-35Nb-7Zr-5Ta-0.06O, at higher temperature ductility will be enhanced (Fig. 6). Finally, Williams et al. [23] have reported similar formation of shear bands and crack propagation along prior β grain boundaries in Ti-10Mo(at. %.) after ω phase precipitation, with these shear bands being absent when this alloy did not contain ω phase precipitates.

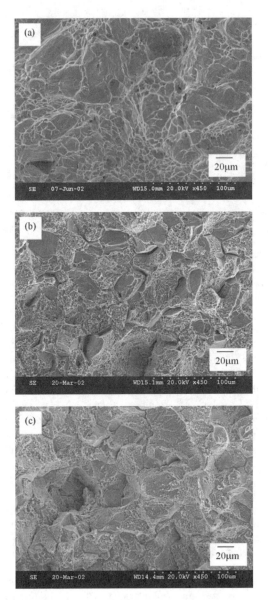

FIG. 16—*Scanning electron micrographs illustrating the tensile fracture surfaces of Ti-35Nb-7Zr-5Ta containing (a) 0.06, (b) 0.46, and (c) 0.68 wt.% oxygen aged at 593°C for 8 h.*

Aging at 482°C for 8 h

As noted previously, increasing the aging temperature for Ti-35Nb-7Zr-5Ta-0.06O did not result in ω and/or α phase precipitation. Thus, aging at 482°C and above should have little effect on either the microstructure or tensile properties of this alloy (Fig. 6).

In contrast, increasing the aging temperature to 482°C tended to promote the formation of fine α, which in turn increased the tensile yield and ultimate tensile strength of the Ti-35Nb-7Zr-5Ta-0.46/0.68O alloys. However, the elongation was reduced. Fractographic evidence suggests that this reduction in tensile ductility can be largely ascribed to increased concentration of plastic deformation of the regions near prior β grain boundaries, i.e., those associated with the precipitate free zones, crack propagation within and along prior grain boundaries, resulting in reduced elongation.

Aging at 538°C for 8 h

A further increase in the aging temperature to 538°C resulted in an increased α precipitation in Ti-35Nb-7Zr-5Ta-0.46/0.68O, as evidenced by an increase in the integrated intensities of α phase peaks (Fig. 12). Although the α precipitates were relatively fine, their precipitation continued to be relatively inhomogeneous, precipitate free zones again being observed along prior β grain boundaries (Fig. 11). Once again, fractography analysis revealed that the crack initiation and propagation in these alloys tended to be restricted to near grain boundary regions.

Aging at 593 °C for 8 h

Finally, aging of Ti-35Nb-7Zr-5Ta-0.46/0.0.68O at 593°C for 8 h resulted in coarsening of the intragranular α phase, widening of the precipitation free zone, and heterogeneous nucleation and growth of α on the prior β grain boundaries (Fig. 14). Fractography analysis revealed that failure once again propagated near or along the prior β grain boundaries. Further evidence for this is also provided by the fine dimpled structure of these regions, these dimples presumably being associated with the grain boundary α phase formed at this high aging temperature.

Conclusions

1. The yield strength of Ti-35Nb-7Zr-5Ta in the solution treated condition increases from 530 to 1081 MPa with increasing oxygen content from 0.06 to 0.68 wt.%, with little apparent effect on tensile elongation.
2. At a fixed oxygen content, low temperature aging results in a further increase in tensile yield strength, with some decrease in tensile elongation, although it still remains at 8 % or higher.
3. Increasing oxygen content suppresses ω phase and promotes α phase formation.
    a. Aging Ti-35Nb-7Zr-5Ta-0.06O at 427°C results in ω phase precipitation; further increase in aging temperature resulted neither in ω nor α phase precipitation.
    b. Aging Ti-35Nb-7Zr-5Ta-0.46O at 427°C results in ω + α precipitation; increasing aging temperature resulted in coarsening of α phase precipitates.
    c. Aging Ti-35Nb-7Zr-5Ta-0.68O results in α phase precipitation; increasing aging temperature resulted in coarsening.

4. Higher temperature aging of Ti-35Nb-7Zr-5Ta-0.46/0.68O additionally results in an increase in the precipitate free zone width and ultimately the heterogeneous precipitation of α on prior β grain boundaries, both tending to result in a reduction in tensile ductility.

References

[1]   Long, M. and Rack, H. J., *Biomaterials*, 19, 1998, pp. 1621–1639.
[2]   Wang, K., *Mat. Sci. Eng. A*, A213, 1996, pp. 134–137.
[3]   Niinomi, M., *Mat. Sci. Eng. A*, A243, 1998, pp. 231–236.
[4]   Niinomi, M., *Metal. Mat. Trans. A*, 33A, 2002, pp. 477–486.
[5]   Niinomi, M., *JOM*, 51, 6, 1999, pp. 32–34.
[6]   Tang, X., Ahmed, T., and Rack, H. J., *J. Mat. Sci.*, 35, 2000, pp. 1805–1811.
[7]   Qazi, J. I. and Rack, H. J., *Medical Device Materials*, S. Shrivastava, Ed., ASM, Cleveland, OH, 2004, pp. 349–356.
[8]   Valiev, R. Z., Stolyarov, V. V., Rack, H. J., and Lowe, T. C., *Medical Device Materials*, S. Shrivastava, Ed., ASM, Cleveland, OH, 2004, pp. 362–367.
[9]   Boyer, R. R., Welsch, G., and Collings, E. W., *Materials Properties Handbook: Titanium Alloys*, ASM, Materials Park, OH, ASM International, 1994, pp. 483–636.
[10] Donachie, Jr., M. J., *Titanium and Titanium Alloys*, ASM, Cleveland, OH, 1982, pp. 33–42.
[11] Ahmed, T. and Rack, H. J., *Mat. Sci. Eng. A*, A243, 1998, pp. 206–211.
[12] Senkov, O. N., Valencia, J., Senkova, S. V., Cavusoglu, M., and Froes, F. H., *Mat. Sci. Tech.*, 2002, pp. 1471–1478.
[13] Kurado, D., Niinomi, M., Morinaga, M., Kato, Y., and Yashiro, T., *Mat. Sci. Eng. A*, A243, 1998, pp. 244–249.
[14] Ahmed, T., Long, M., Silverstri, J., Ruiz, C., Rack, H. J., *Titanium 95: Science and Technology*, P. A. Blenkinsop, W. J. Evans, and H. M. Flower, Eds., IoM, London, UK, 1995, pp. 1760–1767.
[15] Imam, M. A. and Feng, C. R., *Advances in the Science and Technology of Titanium Alloys Processing*, P. J. Bania, D. Eylon, and S. L. Semiatin, Eds., TMS Warrendale, PA, 1997, pp. 435–450.
[16] Hao, Y. L., Niinomi, M., Kuroda, D., Fukunaga, K., Zhou, Y. L., Yang, R., et al., *Met. Mat. Trans. A*, 34A, 2003, pp. 1007–1012.
[17] Ikeda, M., Komatsu, S. Y., Sowa, I., and Niinomi, M., *Met. Mat. Trans.*, 33A, 2002, pp. 487–493.
[18] Kobayashi, E., Doi, H., Yoneyama, T., Hamanaka, H., Gibson, I. R., and Best, S.M., *J. Mat. Sci.: Mat. in Medicine*, 9, 1998, pp. 625–630.
[19] Hao, Y. L., Niinomi, M., Kuroda, D., Fukunaga, K., Zhou, Y.L., Yang, R., et al., *Met. Mat. Trans. A*, 33A, 2002, pp. 3137–3144.
[20] Qazi, J. I., Tsakiris, V., Marquardt, B., and Rack H. J.: *Titanium 2003 Science and Technology*, G. Lutjering and J. Albrecht, Eds., Wiley-VCH Verlag, GmbH & Co. KGaA, Weinheim, Germany, 2004, pp. 1651–1658.
[21] Qazi, J. I., Marquardt, B., Allard, L. F., and Rack, H. J., "Phase Transformations in Ti-35Nb-7Zr-5Ta-(0.06-068)O Alloys," to be published in *Mater. Sci. Eng. C*, 2005.
[22] Williams, J. C., Hickman, B. S., and Marcus, H. L., *Met. Trans.*, 2, 1971, pp. 477–484.
[23] Williams, J. C., Hickman, B. S., and Marcus, H. L., *Met. Trans.*, 2, 1971, pp. 1913–1922.

Journal of ASTM International, October 2005, Vol. 2, No. 9
Paper ID JAI12779
Available online at www.astm.org

Brian Marquardt[1] and Ravi Shetty[2]

Beta Titanium Alloy Processed for High Strength Orthopedic Applications

ABSTRACT: The general material requirements for Ti-15Mo have been standardized for surgical implant applications in ASTM F 2066 [1]. This particular standard is currently limited to one microstructural condition that is produced by beta solution treating and rapidly quenching the material to avoid the formation of alpha phase. When the alloy is processed in this way, it maintains a very high level of ductility but relatively low tensile and bending fatigue strengths. Since beta titanium alloys can be strengthened substantially by thermal processes that introduce alpha phase, it may be possible to adjust the microstructure to achieve a balance of properties that will meet the requirements for more highly stressed applications. Tensile and rotating beam fatigue data for several dual phase material conditions of Ti-15Mo indicate that the alloy can be strengthened substantially without introducing tensile notch sensitivity. However, it was necessary to use an alpha/beta annealing process and avoid beta solution treatment prior to aging at 480 °C to achieve this balance of properties. The potential benefits of alpha/beta processing appear to warrant expansion of ASTM F 2066 beyond the current beta solution treated and quenched condition.

KEYWORDS: Ti-15Mo, beta titanium, fatigue strength, notch sensitivity, tensile properties, microstructure, orthopedic applications, ASTM F 2066

Introduction

Beta titanium alloys are a class of titanium alloys that contain sufficient levels of beta stabilizing elements to inhibit a martensitic phase transformation during cooling. Therefore, they are distinguished from the alpha and alpha/beta classes of titanium alloys by the fact that they can maintain a fully body-centered-cubic (BCC) crystal structure after being quenched from above the beta transus temperature. This retained beta or BCC material condition has often been the focus of study for orthopedic applications since it offers a lower elastic modulus, higher ductility and less notch sensitivity than the widely used Ti-6Al-4V ELI alpha/beta titanium alloy. Commercial beta titanium alloys have been characterized for orthopedic applications [2–3], and a few new beta titanium alloys have been developed specifically for the orthopedics industry [4–6]. Most of these beta titanium alloys have elastic modulus values at or below 75 GPa, ductility near 20 % elongation and notched tensile strength (NTS) to ultimate tensile strength (UTS) ratios near 1.5 for the BCC material condition. Comparatively, Ti-6Al-4V ELI has an elastic modulus of roughly 115 GPa, an ASTM ductility standard of 15 % elongation and a NTS/UTS ratio of approximately 1.4 [2].

Alloys of the beta titanium classification can be thermally processed at intermediate temperatures within the two-phase field (alpha plus beta) to produce a wide range of morphologies with large fractions of hexagonal-close-packed (HCP) alpha phase. In fact,

Manuscript received 19 October 2004; accepted for publication 3 March 2005; published October 2005. Presented at ASTM Symposium on Titanium, Niobium, Zirconium, and Tantalum for Medical and Surgical Applications on 9-10 November 2004 in Washington, DC.
[1] Senior Engineer, Metals Research, Zimmer, Inc., PO Box 708, Warsaw, IN 46581-0708.
[2] Director, Metals Research, Zimmer, Inc., PO Box 708, Warsaw, IN 46581-0708.

material conditions that are quenched to maintain a BCC structure are generally meta-stable for all 'beta' titanium alloys. Thermal processes that promote alpha phase formation increase the elastic modulus and notch sensitivity of beta titanium alloys to levels higher than the meta-stable BCC material condition while reducing the ductility to a level near that of the conventional Ti-6Al-4V ELI alloy. Periodic efforts have been made to identify useful orthopedic beta titanium alloys with lower elastic modulus values [2–6]. In these studies, efforts would correctly focus on the BCC material condition associated with beta solution treatment and rapid quenching to avoid the introduction of alpha phase. Conversely, efforts aimed at the achievement of higher tensile and fatigue strength would likely focus on processes and/or heat treatments within the alpha plus beta phase field. High tensile strength can often be achieved by simply aging beta titanium alloys low in the alpha plus beta phase field, whereas high cycle fatigue strength can be further improved by mechanical processing in the alpha plus beta phase field to refine the prior beta grain size. The challenge lies in the identification of appropriate processing procedures for attaining a good balance between ductility, notch sensitivity, and fatigue strength for highly stressed applications. The identification of such a process for any particular beta titanium alloy may allow for the introduction of stronger, smaller, and/or more fully porous coated titanium orthopedic components. As a result, high strength, multiphase material conditions may broaden the application of beta titanium alloys for orthopedic devices beyond those that benefit from lower elastic modulus values.

Ti-15Mo is a commercially available beta titanium alloy that was initially developed for the chemical industry and was specifically used for its corrosion resistance [7]. The general material requirements for this alloy have been standardized for surgical implant applications in ASTM F 2066 [1]. Ti-15Mo can be produced with cold hearth melting techniques that are capable of producing large, homogeneous ingots. Ti-15Mo is easily processed into round and flat bar forms by standard forging and rolling processes. Like several of the other beta titanium alloys, Ti-15Mo is devoid of elements that are known to be associated with metal allergies. It is also one of the most ductile beta titanium alloys and can be strengthened to a significant degree with relatively short-term aging treatments. Overall, Ti-15Mo is an excellent candidate for the characterization of high strength material conditions with dual phase microstructures of differing morphology and is a possible choice for several high strength orthopedic applications.

Material and Processes

The Ti-15Mo ingot for this project was melted, forged, and rolled at ATI, Allvac. Titanium sponge was blended with pure molybdenum powder to produce compacts for melting a 1360 kg ingot. Since molybdenum is approximately twice as dense as titanium, the elements tend to segregate when melting a primary electrode by conventional vacuum arc remelting (VAR). Therefore, a plasma cold hearth melting process was used to maintain a shallow melt pool and homogeneity during the primary melt. The plasma melted primary ingot measured 430 mm in diameter. A secondary ingot was subsequently melted to 530 mm in diameter by VAR. The results from chemical analysis of the secondary ingot are presented along with the composition limits set by ASTM F 2066 (Table 1). Two values are given for the product analysis when differences were detected between the composition of the top and bottom of the secondary ingot. The beta transus of the ingot was approximately 790 °C.

The double-melted, 530 mm diameter Ti-15Mo ingot was rotary forged to 100 mm diameter billet using a multi-step process. The final reduction step of this process was conducted above the beta transus temperature, and the resultant microstructure was an equiaxed, beta annealed condition. This 100 mm billet material was used to evaluate several rolling processes for the production of round and rectangular bar configurations.

Round bar was beta processed on a continuous rolling mill to a 25 mm diameter, whereas 25 x 75 mm rectangular bar was processed using both beta and alpha/beta conditions on hand mills at ATI, Allvac. These three bar products were given a common thermal treatment after rolling to produce potential high strength material conditions 'A', 'B,' and 'C.' They were first annealed in a vacuum furnace at a temperature high in the alpha/beta phase field. The cooling method from the annealing treatment was a fan-assisted argon gas quench. After the annealing treatment, they were each aged at 480 °C for 4 h. Material condition 'A' (Fig. 1), which was beta processed to 25 mm diameter round bar on a continuous rolling mill prior to thermal treatment, had a banded microstructure with regions of equiaxed prior beta grains and globular alpha grains separated by regions of recovered beta grains and elongated alpha. Material condition 'B' (Fig. 2), which was beta processed to 25 x 75 mm rectangular bar on a hand rolling mill prior to heat treatment, showed little to no evidence of recrystallization. The alpha phase was elongated in some areas but it often appeared in a partially globularized form along variants of the prior beta grains. Material condition 'C' (Fig. 3), which was alpha/beta processed to 25 x 75 mm rectangular bar on a hand rolling mill prior to thermal treatment, had a fully recrystallized and uniformly refined microstructure. The recrystallized prior beta grains and globular alpha of material condition 'C' were roughly equivalent in size to the recrystallized regions in the banded structure of the beta processed round bar, material condition 'A.' The average prior beta grain size was approximately 2 μm, whereas the globular alpha was typically 1 μm or less.

Two additional microstructural conditions were produced from the material that was beta processed on the continuous rolling mill to 25 mm diameter round bar. Material condition 'D' was beta solution treated at 810 °C for 1 h in an air furnace and water quenched to produce an equiaxed beta grain structure 'free' of alpha phase (Fig. 4). The beta grain size of material condition 'D' was approximately 100 μm. Material condition 'E' was produced by using the same beta solution treatment as per material condition 'D' and then aging the material at 480 °C for 4 h in a vacuum furnace. There was no indication of 'feathery' alpha or precipitation free zones along prior beta grain boundaries in material condition 'E' (Fig. 5). Optical microscopy was not sufficient to resolve the size or morphology of any alpha (or omega) second phase particles.

TABLE 1—*Chemical composition of Ti-15Mo bar.*

Element	ASTM F 2066 Limit, weight %	Product Analysis, weight %
Nitrogen	0.05	0.001–0.002
Carbon	0.10	0.006
Hydrogen	0.015	0.0017
Iron	0.10	0.02
Oxygen	0.20	0.15–0.16
Molybdenum	14–16	14.82–15.20
Titanium	balance	balance

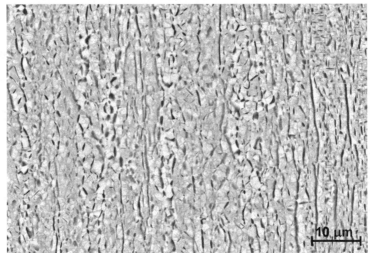

FIG. 1—*Backscattered electron micrograph of material condition 'A'.   Ti-15Mo round bar that was beta processed on a continuous rolling mill. Subsequent to rolling, the material was alpha/beta annealed, argon quenched and then aged at 480 °C for 4 h.*

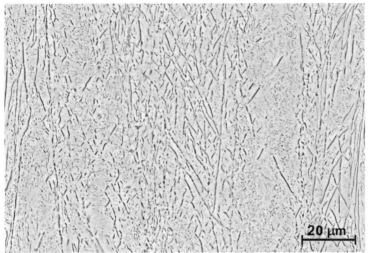

FIG. 2—*Backscattered electron micrograph of material condition 'B'. Ti-15Mo rectangular bar that was beta processed on a hand rolling mill. Subsequent to rolling, the material was alpha/beta annealed, argon quenched and then aged at 480 °C for 4 h.*

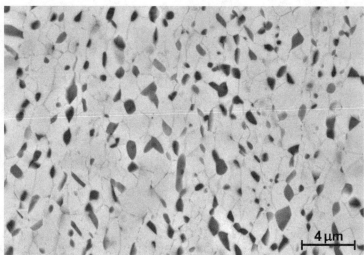

FIG. 3—*Backscattered electron micrograph of material condition 'C'. Ti-15Mo rectangular bar that was alpha/beta processed on a hand rolling mill. Subsequent to rolling, the material was alpha/beta annealed, argon quenched and then aged at 480 °C for 4 h.*

FIG. 4—*Optical micrograph of material condition 'D'. Ti-15Mo round bar that was beta processed on a continuous rolling mill. Subsequent to rolling, the material was beta solution treated and water quenched.*

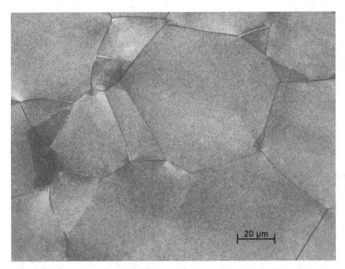

FIG. 5—*Optical micrograph of material condition 'E.' Ti-15Mo round bar that was beta processed on a continuous rolling mill. Subsequent to rolling, the material was beta solution treated, water quenched and then aged at 480 °C for 4 h.*

Test Procedures

Smooth and Notched Tensile

Smooth and notched tensile specimens were machined and tested at Metcut Research. The smooth test specimen configuration had nominal gage dimensions of 6.35 mm diameter by 34.5 mm length. Smooth tensile tests were conducted in accordance to ASTM E-8 with a strain rate of 0.005 per min through the 0.2 % yield strength and a head rate of 1.3 mm per min to failure. The notched tensile specimen configuration had a nominal notch diameter of 6.35 mm with a stress concentration factor (K_t) of 3.2. The notched tensile tests were conducted in accordance to ASTM E-8 with a head rate of 0.76 mm per min to failure.

Rotating Beam Fatigue

Rotating beam fatigue specimens were machined at Metcut Research and tested at Zimmer. The specimen configuration had a nominal gage diameter of 4.76 mm. The R ratio of the test was -1 and the frequency was 50 hertz.

Results and Discussion

Smooth Tensile

Processing parameters and associated smooth bar tensile properties are reported for material conditions 'A' through 'E' (Table 2). Each value represents the average of two test results for specimens with the tensile axis parallel to the rolling direction. The data for material conditions 'A', 'B,' and 'C' in Table 2 represent different rolling processes with a common annealing and aging treatment conducted below the beta transus temperature. The ultimate strength values for these three material conditions range from 1280–1320 MPa with the round bar from the continuous rolling mill, material condition 'A', having the lowest strength. The highest strength for the three conditions was associated with the material that was processed below the beta transus temperature, material condition 'C'. The microstructure for the material rolled below the beta transus temperature (Fig. 3) was fully recrystallized and uniformly refined. This uniform microstructural refinement may partially account for the higher strength value. However, unresolved precipitates and crystallographic texture must also be taken into account. The ductility values for these three material conditions in Table 2 range from 9–14 % elongation with the highest value being associated with the material from the continuous rolling mill, material condition 'A', rather than the uniformly refined material that was processed below the beta transus temperature, material condition 'C.' This unexpected result may indicate that crystallographic texture has influenced the strength and ductility values. The lower strength and higher ductility for the round bar from the continuous mill would be consistent with a stronger rolling texture that may develop as a result of increased total strain during the rolling process and non-uniform or incomplete recrystallization as shown previously (Fig. 1).

Material conditions 'D' and 'E' were tested after solution treatment above the beta transus temperature (Table 2). Material condition 'D' was tested without aging whereas material condition 'E' was aged at 480 °C for 4 h. The material that was tested without aging had an ultimate strength of only 770 MPa but ductility values near 40 % elongation and 80 % reduction of area. These high ductility values are of particular interest for orthopedic trauma applications [2] but the low tensile strength limits the opportunity for a wider range of applications. The aged material condition had an ultimate strength above 1400 MPa but ductility values of only 2 % elongation and 5 % reduction of area. Ductility values as low as this are not promising and essentially eliminate this processing approach from consideration for high strength orthopedic applications. Higher aging temperatures may offer a better balance between strength and ductility for beta solution treated Ti-15Mo. In addition, there are other beta titanium alloys that maintain a better balance between strength and ductility after beta solution treatment and aging [8] that may offer greater potential for this particular processing approach.

Most tensile data for orthopedic beta titanium alloys in the open literature are associated with beta rolling processes and beta annealing cycles without aging [2,4]. In addition, many beta titanium alloys have higher tensile strength than Ti-15Mo in the beta solution treated and rapidly quenched condition. Therefore, beta titanium alloys with higher tensile strength in the BCC material condition, such as TMZF [4], can be used for higher stress applications without aging. As such, the benefits associated with alpha/beta processing and aging may be more significant for Ti-15Mo than for many of the other beta titanium alloys. However, tensile test results for alpha/beta annealed versus beta solution treated Ti-15Mo-5Zr-3Al [9] are comparable to the Table 2 values for Ti-15Mo. The test results for both Ti-15Mo and Ti-15Mo-5Zr-3Al indicate that the high strength material conditions, which are produced by aging low in the alpha plus beta phase field, maintain a better balance of ductility when the bar products are annealed below the beta transus temperature.

TABLE 2—*Smooth tensile properties and process conditions for Ti-15Mo bar.*

ID	Bar Dimensions	Rolling Process	Anneal	Age °C	UTS MPa	0.2 % YS MPa	Elong. %	RA %
A	25 mm diameter	β, Continuous Mill	α/β + ArQ	480	1280	1210	14	59
B	25x75 mm	β, Hand Mill	α/β + ArQ	480	1290	1240	9	32
C	25x75 mm	α / β , Hand Mill	α/β + ArQ	480	1320	1290	9	32
D	25 mm diameter	β, Continuous Mill	β + WQ	NA	770	610	38	80
E	25 mm diameter	β, Continuous Mill	β + WQ	480	1420	1370	2	5

ArQ = argon quench in a vacuum furnace, WQ = water quench from an air furnace

Notched Tensile

Notched tensile data for Ti-15Mo with a K_t value of 3.2 are reported (Table 3) along with the associated processing parameters for material conditions A through E. The data have been reported in ratios that have been published for other titanium biomaterials [2]. The ratio of the notched tensile strength (NTS) divided by the ultimate tensile strength (UTS) is a relative index of notch sensitivity where values below 1.1 represent notch sensitive materials. The notch strength ratio (NSR) is calculated by dividing the NTS by the 0.2 % yield strength. The NSR is generally thought to give a relative indication of plane-strain fracture toughness [2]. Each ratio represents an average result after testing two notched specimens and two smooth specimens per material condition.

The notched tensile data for the alpha/beta annealed material conditions 'A' through 'C' are tightly grouped with values ranging from 1.32–1.36 for NTS/UTS and 1.36–1.43 for NSR. Therefore, none of these conditions are notch sensitive under tensile loading conditions. The highest ratios for these three conditions were associated with the round bar from the continuous rolling mill, material condition 'A.' Comparison with the smooth tensile data indicates that the most ductile condition also had the highest notch sensitivity ratios.

The beta solution treated material conditions 'D' and 'E' have divergent notch sensitivity values that are consistent with the ductility values for smooth bar tensile testing. The material with low ductility that was aged after beta solution treatment, material condition 'E', was the only notch sensitive material with a NTS/UTS value of 0.74. Material condition 'D', which had low strength and high ductility values for smooth bar tensile testing, had the highest overall notch sensitivity ratios. However, the notch sensitivity ratios for these large cross section bar materials were not as high as reported for thin cross sections of a similar material condition (NSR = 2.46 for beta solution treated and quenched Ti-15Mo) [2]. Further coordinated analysis would be required to determine if this is due to differing test procedures, crystallographic texture, grain size, or other experimental parameters.

The smooth tensile data indicated that the aged material conditions of Ti-15Mo did not work harden as much as material condition 'D' which was beta solution treated and water quenched. As a result, the UTS and 0.2 % yield strengths were similar for the aged materials thus causing the NTS/UTS and NSR ratios to be similar as well. Therefore, the aged Ti-15Mo NTS/UTS ratios for notch sensitivity are similar to those reported for other titanium biomaterials while the NSR values for aged Ti-15Mo are somewhat lower [2].

TABLE 3—*Notched tensile ratios for Ti-15Mo bar.*

ID	Bar Dimensions	Rolling Process	Anneal	Age, °C	NTS/UTS	NSR
A	25 mm diameter	β, Continuous Mill	α/β + ArQ	480	1.36	1.43
B	25x75 mm	β, Hand Mill	α/β + ArQ	480	1.34	1.39
C	25x75 mm	α / β , Hand Mill	α/β + ArQ	480	1.32	1.36
D	25 mm diameter	β, Continuous Mill	β + WQ	NA	1.46	1.84
E	25 mm diameter	β, Continuous Mill	β + WQ	480	0.74	0.77

ArQ = argon quench in a vacuum furnace, WQ = water quench from an air furnace.

Rotating Beam Fatigue

Rotating beam fatigue data are shown for material conditions 'A' through 'C' (Fig. 6). Although the smooth and notched tensile test results were tightly grouped for each of these alpha/beta annealed and aged conditions, the fatigue test data show significantly different results. Material condition 'B', with a recovered rather than recrystallized microstructure (Fig. 2) after beta processing to 25 x 75 mm flat bar on hand rolling mills, had a substantially lower fatigue strength than material conditions 'A' and 'C.' It is likely that the refined prior beta grain size associated with the partially recrystallized microstructure of material condition 'A' (Fig. 1) and the fully recrystallized microstructure of material condition 'C' (Fig. 3) accounted for an increased resistance to crack nucleation and improved fatigue strength. The fatigue strength of titanium alloys is often dictated by factors such as prior beta grain size, grain boundary alpha and/or crystallographic texture since these parameters influence crack initiation. Previously reported results for Ti-15Mo-5Zr-3Al [9] also indicated that prior beta grain size was an influential factor in determining the rotating beam fatigue strength of this similar beta titanium alloy.

The fatigue test data (Fig. 6) also show a larger degree of scatter for material condition 'A' than for material conditions 'B' and 'C'. The results for material condition 'B' are representative of a recovered microstructure, and the results for material condition 'C' are representative of a fully recrystallized microstructure. Material condition 'A' had a banded microstructure with both recrystallized and recovered regions (Fig. 1). Therefore, it is likely that there were variations in the volume fraction of recovered versus recrystallized grains from one test specimen to another for material condition 'A'. This variation in microstructure may account for the increased scatter for the fatigue test data associated with material condition 'A'. If so, it may be useful to modify the rolling practice on the continuous rolling mill to produce a fully recrystallized microstructure rather than a partially recrystallized microstructure after subsequent heat treatment within the alpha plus beta phase field.

The rotating beam fatigue data for material condition 'C' has been compared to two other titanium alloys (Fig. 7). The Ti-6Al-4V ELI material for this comparison was heat-treated high in the alpha plus beta phase field and then quenched with fan assisted argon gas in a manner similar to that used for the Ti-15Mo test specimens. The annealing temperature was raised to account for the higher beta transus temperature for Ti-6Al-4V ELI. The microstructure of the Ti-6Al-4V ELI (Fig. 8) consisted of equiaxed or nearly spherodized alpha grains with a width of approximately 15–20 μm. A small fraction of beta and acicular alpha remained between equiaxed alpha grains. The Ti-12Mo-6Zr-2Fe fatigue data are from the work of Wang et al. [4] that evaluated the beta solution treated condition after 'rapid cooling.' The beta grain size for the Ti-12Mo-6Zr-2Fe was similar to material conditions 'D' and 'E' at approximately 100 μm. The testing of Ti-12Mo-6Zr-2Fe was conducted at a higher frequency of approximately 167 hertz.

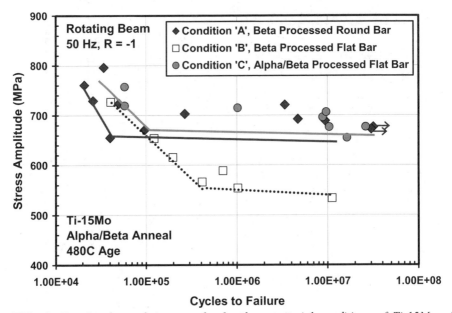

FIG. 6—*Rotating beam fatigue results for three material conditions of Ti-15Mo.   The microstructures associated with the test results are shown in Figs. 1–3.*

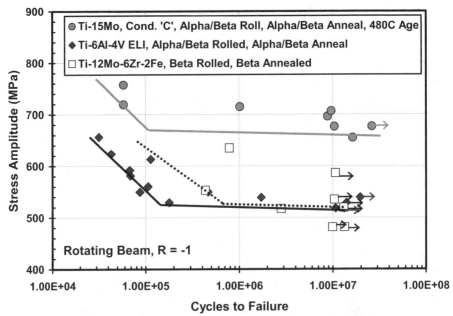

FIG. 7—*Rotating beam fatigue results for three titanium alloys.*

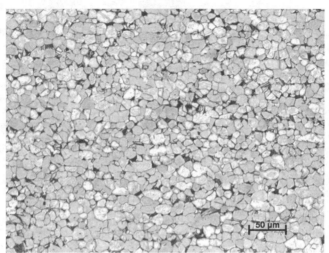

FIG. 8—*Optical micrograph of Ti-6Al-4V ELI bar material that was rolled and then annealed high in the alpha plus beta phase field.*

The fatigue strength of Ti-6Al-4V ELI and Ti-12Mo-6Zr-2Fe were roughly equivalent. This result is consistent with the Wang et al. results, which compared Ti-12Mo-6Zr-2Fe to mill annealed Ti-6Al-4V [4]. The refined and strengthen version of Ti-15Mo represented by material condition 'C' has a substantially higher fatigue strength than each of the other alloys (Fig. 7). However, the fatigue data shown previously (Fig. 6) for material condition 'B,' with a recovered rather than recrystallized microstructure, was roughly equivalent to the fatigue data for the two additional titanium alloys. Refinement of the prior beta grain size was necessary in order to show a rotating beam fatigue strength advantage over Ti-6Al-4V ELI and Ti-12Mo-6Zr-2Fe.

Summary and Conclusions

A broad range of microstructural conditions can be produced for the Ti-15Mo beta titanium alloy by altering the rolling and heat treatment procedures. Beta solution treated and water quenched material has very high ductility and relatively low strength while direct aging the same material leads to high strength and low ductility. When the rolled materials are annealed in the alpha plus beta phase field and aged at 480 °C, a good balance of tensile and notched tensile properties are achieved. The overall balance of properties that was achieved for this recrystallized dual phase microstructure warrants consideration for expanding ASTM F 2066 to include a larger range of microstructural conditions.

Acknowledgments

The authors are grateful to Sushil Bhambri and Eric Langley of Zimmer for their assistance with scanning electron microscopy and rotating beam fatigue testing, respectively. In addition, the authors would like to thank Julie Bock and Kai Lorcharoensery for assistance with optical microscopy and acknowledge Mike Hawkins and Cheryl Blanchard for their support of this project.

References

[1] ASTM F 2066, "Standard Specification for Wrought Titanium-15Molybdenum Alloy for Surgical Implant Applications (UNS R58150)," *Annual Book of ASTM Standards*, ASTM International, West Conshohocken, PA.

[2] Disegi, J. and Zardiackas, L., "Metallurgical Features of Ti-15Mo Beta Titanium Alloy for Orthopedic Trauma Applications," *Medical Device Materials, 8–10 September 2003*, ASM International, Materials Park, OH (2004) pp. 337–342.

[3] Steinemann, S. G., Mausli, S., Szmukler-Moncler, S., Semlitsch, M., Pohler, O., Hintermann, H. E., and Perren, S. M., "Beta-Titanium Alloy for Surgical Implants," *Titanium '92, Science and Technology*, TMS, Warrendale, PA, 1993, pp. 2689–2696.

[4] Wang, K., Gustavson, L., and Dumbleton, J., "The Characterization of Ti-12Mo-6Zr-2Fe – A New Biocompatible Titanium Alloy Developed for Surgical Implants," *Beta Titanium Alloys in the 1990s*, TMS/AIME, Warrendale, PA, 1993, pp. 49–60.

[5] Niinomi, M., Hattori, T., Morikawa, K., Kasuga, T., Suzuki, A., Fukui, H., and Niwa, S., "Development of Low Rigidity Beta-type Titanium Alloy for Biomedical Applications," *Materials Transactions*, Vol. 43, No. 12, 2002, pp. 2970–2977.

[6] Ahmed, T., Long, M., Silvestri, J., Ruiz, C., and Rack, H. J., "A New Low Modulus, Biocompatible Titanium Alloy," *Titanium '95, Science and Technology*, IoM, London, UK, 1995, pp. 1760–1767.

[7] IMI Titanium 205, Alloy Data Sheet, IMI Titanium Limited, Birmingham, England.

[8] Qazi, J. I., Tsakiris, V., Marquardt, B., and Rack, H. J., "The Effect of Duplex Aging on the Tensile Behavior of Ti-35Nb-7Zr-5Ta-(0.06-0.7)O Alloys," *Ti-2003, Science and Technology*, Wiley-VCH, Weinheim, Germany, 2004, pp. 1651–1658.

[9] Tokaji, K., Bian, J. C., Ogawa, T., and Nakajima, M., "The Microstructure Dependence of Fatigue Behavior in Ti-15Mo-5Zr-3Al Alloy," *Materials Science and Engineering*, A213 1996, pp. 86–92.

Journal of ASTM International, September 2005, Vol. 2, No. 8
Paper ID JAI13033
Available online at www.astm.org

Victor R. Jablokov,[1] *Michael J. Nutt,*[2] *Marc E. Richelsoph,*[2] *and Howard L. Freese*[3]

The Application of Ti-15Mo Beta Titanium Alloy in High Strength Structural Orthopaedic Applications

ABSTRACT: Titanium-15Molybdenum beta titanium alloy is a generic alloy that has been around for a long time, but has never found wide use in the aerospace industry. Beta titanium alloys, in general, and corrosion-resistant biocompatible titanium alloys, in particular, attracted a lot of attention in the medical and surgical device industry in the late 1980s and early 1990s. Until that time, all of the titanium grades used for medical implant applications were based on ASTM "F" standards modeled from aerospace industry and AMS specifications. The CP titanium grades (CP-1, CP-2, and CP-4) and two $\alpha + \beta$ alloys, Ti-6Al-4V ELI and Ti-6Al-4V, were widely used in aerospace applications and found early acceptance to become the reference metallic materials in ASTM F 748. Two important factors held Ti-15Mo back from commercial applications and use: the physical metallurgy of the alloy and the inability of the then-current reactive metals melting and processing machinery to handle this unusual binary alloy. This paper will briefly review high-technology manufacturing advances and processing technology that have enabled the reliable production of large-scale lots of Ti-15Mo bar and rod product forms. Additionally, mechanical and metallurgical data for semi-finished high strength rod product suitable for highly stressed orthopaedic implant applications will be presented.

KEYWORDS: metastable beta titanium, metal, orthopaedic, Ti-15Mo, high strength, titanium alloys, titanium, cold work, cold reduction, aging, aging response, heat treatment, spinal devices

Introduction

The development, basic metallurgy, mechanical and corrosion properties, and biocompatibility of Ti-15Mo alloy have been described sufficiently as to now appear as references in a material supplier's technical data sheet [1], with the binary alloy composition dating back to the 1950s. With the biomedical interest in beta titanium alloys surfacing in the 1990s, publication of the characteristics of Ti-15Mo, most notably by Zardiackas, Mitchell, and Disegi in 1996 [2]; Bogan, Zardiackas, and Disegi in 2001 [3]; and Disegi and Zardiackas in 2003 [4], has laid a strong foundation of information before the medical device manufacturers for consideration in the design of their products. Sufficient biocompatibility test data, as well as chemical, mechanical, and metallurgical data, have permitted creation of an ASTM material standard for wrought Ti-15Mo alloy for surgical implant applications, which was published in 2000 and designated F 2066 [5].

One of the reasons for the interest in beta titanium alloys in general and Ti-15Mo specifically is the lower modulus of elasticity, or lower stiffness. Until very recently, the titanium alloys used for high strength orthopaedic applications have been limited to alpha-beta alloys. Alpha-

Manuscript received 25 October 2004; accepted for publication 18 March 2005; published September 2005.
Presented at ASTM Symposium on Titanium, Niobium, Zirconium, and Tantalum for Medical and Surgical Applications on 9-10 November 2004 in Washington, DC.
[1] Research & Development, ATI Allvac, an Allegheny Technologies company, Monroe, NC 28110 USA.
[2] Spinal Innovations, Inc., Bartlett, TN 38133 USA.
[3] Business Development, ATI Allvac, an Allegheny Technologies company, Monroe, NC 28110 USA.

beta alloys like Ti-6Al-4V (ASTM F 1472), Ti-6Al-4V ELI (ASTM F 136), and Ti-6Al-7Nb (ASTM F 1295) are the standards for biocompatibility, corrosion resistance, and fatigue strength, with annealed yield strength minimums of 795 to 860 MPa (115 to 125 ksi) [6,7,8]. However, there are orthopaedic applications that could benefit from a lower modulus of elasticity than those provided by the alpha-beta titanium alloys. In the early 1990s, beta titanium alloys had this potential in the solution annealed condition, depending on composition, so medical device manufacturers had to choose between commercially available low modulus alloys that needed biocompatibility testing or private development of new biocompatible alloys that would meet their low modulus requirements. At least two such proprietary alloys proceeded to actual application in orthopaedic devices, which received FDA approval for marketing and eventually became ASTM implant material specifications. Those two proprietary alloys are ASTM F 1713 (Ti-13Nb-13Zr), published in 1996, and ASTM F 1813 (Ti-12Mo-6Zr-2Fe), published in 1997, having annealed minimum yield strengths of 345 and 897 MPa (50 and 130 ksi), respectively [9,10]. The non-proprietary beta titanium alloy chosen for orthopaedic implant applications became the ASTM F 2066 standard (Ti-15Mo), with a minimum annealed yield strength of 483 MPa (70 ksi) [5]. Table 1 gives a comparison of the modulus of elasticity and the ASTM specification minimum mechanical properties for each of these alloys in the annealed condition.

TABLE 1—*Mechanical properties for selected ASTM material specifications.*

Alloy Designation	Condition / Microstructure	Elastic Modulus, GPa (msi)	ASTM Standard	Ultimate Tensile Strength, min. MPa (ksi)	Yield Strength (0.2 % offset), min. MPa (ksi)	Elongation in 4D, min. %	Reduction of Area, min. %
Ti-6Al-4V ELI	Annealed / Alpha + Beta	98.4 (14.3) [2]	F 136	860 (125)	795 (115)	10	25
Ti-6Al-4V	Annealed / Alpha + Beta	110 (16) [11]	F 1472	930 (135)	860 (125)	10	25
Ti-6Al-7Nb	Annealed / Alpha + Beta	99.9 (14.5) [2]	F 1295	900 (130.5)	800 (116)	10	25
Ti-15Mo	Annealed / Beta	77.7 (11.3) [2]	F 2066	690 (100)	483 (70)	20	60
Ti-13Nb-13Zr	Annealed / Beta	64–77 (9.3–11.2) [12]	F 1713	550 (80)	345 (50)	15	30
Ti-12Mo-6Zr-2Fe	Annealed / Beta	74–85 (10.7–12.3) [13]	F 1813	931.5 (135)	897 (130)	12	30

The Need for High-Strength Titanium Alloys in Orthopaedic Applications

There are orthopaedic implant applications that are more concerned with higher strength than lower modulus. In orthopaedic applications where physical size or higher stiffness, or both, are dominating design features, a biocompatible metallic material is needed that has substantially higher strength than annealed metallic materials can provide. In the past, one might be forced to consider a high-strength, cold-worked stainless steel implant alloy or a cold-worked and age-hardened cobalt-base implant alloy. The users, physicians performing the surgical procedures, prefer titanium alloys, not only for the superior biocompatibility and corrosion resistance, but also for the clarity of magnetic resonance and radiographic images of implanted devices. Although several orthopaedic industry segments may find applications for a high strength, moderate stiffness, titanium implant alloy (such as a component in a total knee system, a component in a total hip system, a substrate for a high-hardness articulation coating, or a very high strength pin or wire), a definite field of application exists in the spinal fixation segment of the industry.

A high-strength titanium material permits the design of lower profile, smaller components for spinal fixation. Benefits include: less protrusion of the implanted device, smaller incisions or minimally invasive surgical techniques, and less vertebral bone removal or contouring, without compromising spinal construct strength and rigidity. Spinal rods, connectors, screws, hooks, plates, and discs are all possible high-strength titanium alloy candidates. The photographs displayed in this paper depict the evolution of spinal fixation implant design for one company, using ASTM F 136 (Ti-6Al-4V ELI) almost exclusively for all components of every implant system (Fig. 1). The size of the components and the resulting implant profile have been reduced dramatically with each iteration, while maintaining or increasing the accepted competitive standard for component and construct static and fatigue strength, using accepted spinal device test methods described in ASTM F 1798, F 2193, and F 1717 (see Fig. 2) [14,15,16]. Anticipating that spinal fixation devices will continue to evolve to even smaller dimensions, the titanium alloy that will be used must be substantially stronger than ASTM F 136, or age hardened Ti-6Al-4V (ASTM F 1472). The highest strength possible, with acceptable ductility, from Ti-15Mo (ASTM F 2066) became the objective for this product and process development program.

Beta titanium alloys, as well as alpha-beta titanium alloys, can achieve increases in strength by various mechanical and thermal treatments. In general, beta titanium alloys have a more robust response to thermal processing, or cold work and thermal processing, than alpha-beta titanium alloys. The initial goal, knowing the capability of some commercial beta titanium alloys, was to roughly double the yield strength available from the popular alpha-beta implant alloy ASTM F 136 (Ti-6Al-4V ELI). Therefore, the specification "aims" became 1379 MPa (200 ksi) minimum yield strength, 1448 MPa (210 ksi) minimum tensile strength, 6 % minimum elongation, and 18 % minimum reduction of area. The highest strength thermal treatments available for the more responsive and popular alpha-beta alloy, F 1472 (Ti-6Al-4V), could not come close to these goals. When considering the proven biocompatible beta titanium alloys available for process development to maximize strength, based upon the information in their respective ASTM standards, there were only three to choose from, and only one of them, Ti-15Mo (ASTM F 2066), was not a proprietary alloy assigned to a medical device manufacturer. As stated above, others have contributed an excellent foundation of research and test results for Ti-15Mo alloy in the annealed condition. This paper describes the development work and results achieved in obtaining high strength Ti-15Mo alloy bar and rod products suitable for highly stressed orthopaedic implant applications.

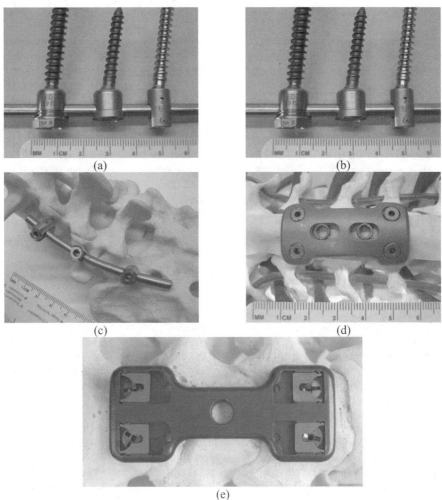

FIG. 1—(a) *Three different polyaxial screw assemblies from one company, mounted on a
5.5 mm rod, demonstrating evolution of fixation systems to smaller component
size.*
(b) *Two different polyaxial screw assemblies compared to a fixed screw, and two
different hook body designs, illustrating differences in component size.*
(c) *Three polyaxial screw assemblies and a contoured 5.5 mm rod connecting L 4,
L 5, and the sacrum on one side of a spine model, demonstrating the
advantages of smaller device dimensions.*
(d) *Cervical plate attached to cervical spine model, connecting C 4 to C 6,
illustrating conventional size and design features.*
(e) *Newer cervical plate, from the same company, attached to cervical spine
model to show reduced dimensions and improved design features.*

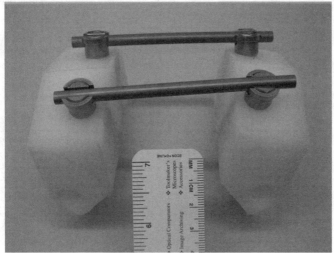

FIG. 2—*Polyaxial screw and rod assemblies after fatigue testing in accordance with the Lumbar Bilateral Construct Test Setup specified in ASTM F 1717.*

Manufacturing High Strength Ti-15Mo Rod

A schematic of the production processes used to produce high strength Ti-15Mo rod stock is shown in Fig. 3. Production is initiated by blending raw materials in the form of titanium sponge and molybdenum master alloy and pressing the loose material into "compacts" or "pucks" that are used as input for Plasma Arc Melting (PAM). PAM is the preferred primary melt process, since unlike primary VAR melting, the need to consolidate the raw material into a single electrode prior to melting is eliminated. Additionally, PAM melting is superior to VAR in eliminating both high density inclusions and hard alpha defects [17], and for alloys that contain elements that are prone to segregation during melting (such as molybdenum, tantalum, or tungsten) the feasibility of uniformly alloying the material is increased as compared to VAR melting. Following the creation of the initial consolidated electrode by PAM melting, Vacuum Arc Remelting (VAR) is used to remelt the electrode and provides additional refinement and minimizes inhomogeneities, resulting in an internal ingot structure more conducive to subsequent hot working.

Following ingot conditioning (i.e., machining) and inspection (to verify ingot chemistry), the ingot is press forged to break down the cast structure and produce a smaller intermediate billet in preparation for rotary forging. Rotary forging is used to break down the structure further and to produce a reroll billet that can be continuously rolled to coil. In the rotary forging operation, the billet is held in rotating jaws that move laterally through four hammers that simultaneously forge the material. In contrast to pressing, this high productivity forging operation generally provides better control over the hot working parameters and results in a more uniform wrought structure [18]. After rotary forging, the billet is again conditioned and ultrasonically inspected to ensure the absence of defects then macro inspected to evaluate the as-rotary forged structure. Once the inspection of the reroll billet is completed, the reroll billet is shipped to a continuous high volume rolling mill for hot rolling to coil.

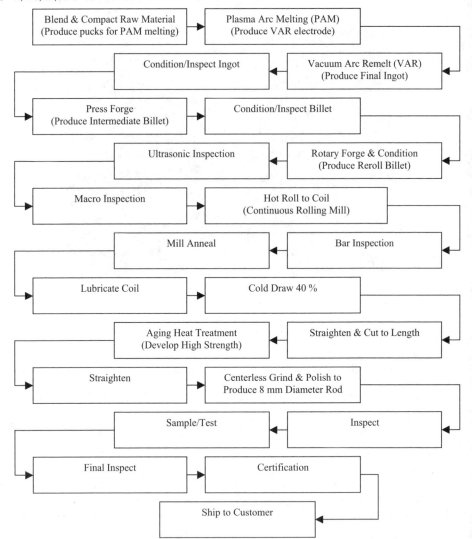

FIG. 3—*Process outline for production of Ti-15Mo high strength rod.*

Following coil rolling, the coil is inspected to ensure that the diameter is within required tolerances for further processing. Mill annealing is performed to solution and recrystallize the material in preparation for cold drawing. In addition to inducing work into the material as the cross-sectional area is reduced during drawing, which aids subsequent recrystallization during heat treating and leads to a finer grain size, cold drawing also acts as the initial straightening operation. This facilitates the post-drawing straightening operations necessary to produce straight rod. When the material has been straightened adequately, the rods are cut to the necessary length and aged to develop high strength properties. If warping occurs as a result of the aging thermal treatment, the straightening operation is repeated. Centerless grinding and rod polishing are the final processes used to finish the material and to achieve the necessary diameter tolerances set by the customer. For further detail regarding the production practices used to produce high strength Ti-15Mo alloy rod stock and other titanium alloys for the biomedical industry, see Davis et al. [18].

Materials and Experimental Procedure

The goals of this study were to characterize the effects of varying cold work reduction, aging time and temperature on the mechanical properties of Ti-15Mo rod. Results are presented from two testing programs; an earlier program performed to characterize the effect of varying cold work reduction on the properties of Ti-15Mo [19] and a recent study performed to evaluate the effect of aging time and temperature on Ti-15Mo that was cold drawn 40 %.

To characterize the effect of cold reduction on the strength and ductility of Ti-15Mo alloy, 8 mm diameter Ti-15Mo rod in the solution annealed condition was-drawn using reductions of 15 %, 25 %, and 40 %. The 8 mm diameter rod was initially produced by hot rolling a reroll billet to 12.7 mm diameter rod; cold drawing to 8.5 mm to reduce the cross-section; annealing at 802°C for 30 min; and water quenching (to recrystallize the microstructure), cold straightening, and centerless grinding to 8 mm to remove alpha case and surface defects. Portions of both the solution treated and cold drawn rods were aged at 482°C and 538°C for 4 h and air-cooled using a standard atmosphere laboratory furnace prior to testing. One room temperature tensile test was performed to characterize the mechanical properties at each condition per ASTM E 8 requirements using specimens with gage diameters of 6.350 mm ± 0.127 mm (0.250 in. ± 0.005 in.).

The test material used to characterize the effect of aging time and temperature on 40 % cold worked Ti-15Mo rod was produced as follows. Previously produced solution treated 8 mm diameter machined straight rod was heat tinted at 802°C for 15 min and water quenched to develop an oxide layer on the surface of the material that would aid the adherence of cold drawing lubricant. The straight rod was then lubricated and cold drawn 40 % to a diameter of 6 mm. Immediately following cold drawing, several rods were diverted and used for evaluating the aging response of Ti-15Mo cold drawn rod. The remaining production material was cold straightened, aged, cold straightened again, and finally centerless ground to produce the 5.4 mm high strength rod shown in Fig. 4.

FIG. 4—*High strength 5.4 mm diameter Ti-15Mo 40 % cold drawn, fully heat-treated rod stock.*

Four aging temperatures were selected to characterize the aging response of 40 % cold drawn Ti-15Mo rod in this study, 454°C, 482°C, 510°C, and 538°C. Additionally, hold times of 1, 2, 4, 8, 12, 16, and 24 h were used to generate aging curves at each temperature of interest. For each temperature and time combination, two room temperature tensile tests were performed per ASTM E 8 requirements using specimens with gage diameters of 4.064 mm ± 0.076 mm (0.160 in. ± 0.003 in.). Optical and scanning electron microscopy was performed to characterize the microstructure and morphology of the fracture surfaces, respectively.

Results and Discussion

Table 2 and Figs. 5 and 6 reveal the tensile properties of Ti-15Mo rod after various heat treatment conditions and cold reduction levels were applied. These data were generated previously as part of a study to establish the optimum cold work reduction level for Ti-15Mo high strength rod [19]. Figure 5a shows that for Ti-15Mo rod that has not been aged, the ultimate tensile strength (UTS) increases as the level of cold reduction increases. However, after aging, the material is strengthened to the same level regardless of the amount of cold work reduction induced into the material. Additionally, aging at 538°C overages the material, and the strength decreases ~20 % as compared to material aged at 482°C. In Fig. 5b, the yield strength (YS) for the various test conditions revealed no increase in YS with increasing cold reduction; however, the post aging trends matched those observed in Fig. 5a.

The ductility trends for the various conditions tested are shown in Figs. 6a and 6b. For material that was not age hardened, elongation was observed to decrease with increasing cold reduction, but the reduction of area remained relatively constant. Post aging, ductility was observed to increase with increasing cold reduction. Note that the low YS (~554 MPa) and high UTS (~758 MPa), as well as high percent elongation (~45 %) in the solution treated and unaged condition, are reflective of the fact that this alloy deforms by twinning. Other metastable beta titanium alloys that deform by uniform slip have higher YS values and lower elongation values in the solution treated condition. For example, the YS and percent elongation for solution treated Ti-12Mo-6Zr-2Fe (TMZF® beta titanium alloy) are approximately 1000 MPa and 20 %, respectively.

TABLE 2—*Mechanical properties of solution treated; solution treated and aged; solution treated and cold drawn; and solution treated, cold drawn, then aged Ti-15Mo rod [19].*

Cold Work (%)	Temp. (°C)	Aging Time (h)	U.T.S. (MPa)	Y.S. (MPa)	Elong. (%)	R.O.A. (%)
	N/a	N/a	758	554	45.8	72.7
0 %	482	4	1426	1345	7.4	6.7
	538	4	1129	1031	9.6	17.1
	N/a	N/a	949	530	21.9	75.5
15 %	482	4	1348	1240	3.1	6.7
	538	4	1094	1002	9.4	26.9
	N/a	N/a	1097	565	18.8	73.0
25 %	482	4	1348	1269	4.7	18.8
	538	4	1128	1043	9.4	37.3
	N/a	N/a	1256	495	15.8	70.4
40 %	482	4	1402	1303	6.3	25.1
	538	4	1104	1027	14.1	56.1

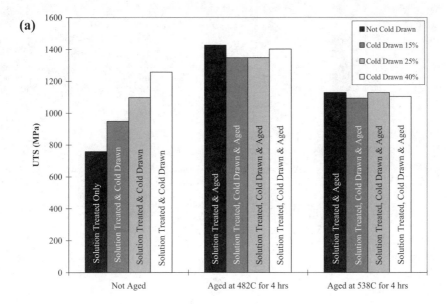

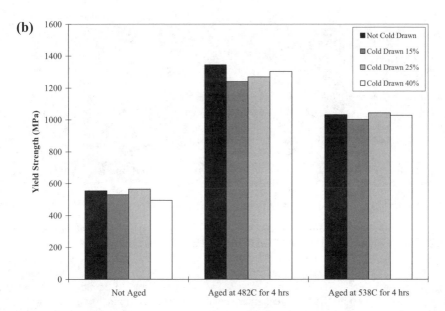

FIG. 5—*Ultimate tensile strength (*a*) and yield strength (*b*) of solution treated; solution treated and aged; solution treated and cold drawn; and solution treated, cold drawn, then aged Ti-15Mo rod [19].*

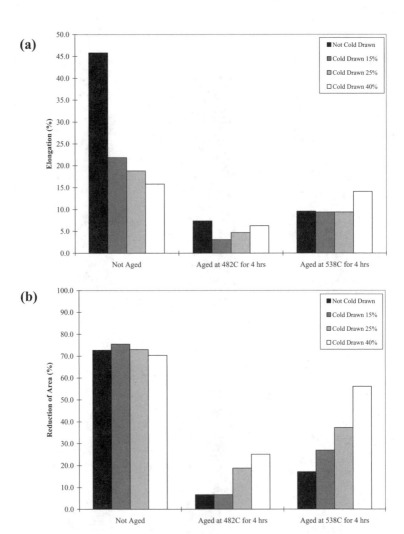

FIG. 6—*Elongation* (a) *and reduction of area* (b) *of solution treated; solution treated and aged; solution treated and cold drawn; and solution treated, cold drawn, then aged Ti-15Mo rod (see Fig. 5a) [19].*

In summary, prior to age hardening, the UTS increased, but YS remained constant with increasing cold reduction. After age hardening, the material strengthened to the same level regardless of the magnitude of cold reduction. In contrast, post aging, the ductility improved with increasing cold reduction. Therefore, since the best combination of strength and ductility was observed in age hardened 40 % cold drawn Ti-15Mo rod, 40 % was chosen as the optimum level of cold reduction for further study.

Table 3 and Figs. 7–10 depict the results of the aging experiments performed on Ti-15Mo 40 % cold drawn rod. Solution treated and as-drawn properties are shown on all of the figures for comparison. The dependence of aging temperature on UTS and YS strength of 40 % cold drawn rod is shown on Figs. 7 and 8. It is clearly shown that as the aging temperature increases, both the UTS and YS decrease. Additionally, an extremely fast aging response is observed in this material where the highest strength is observed at the shortest aging time, and as the hold time increases from 1 to 24 h, the strength of the material decreases.

It is well known that for metastable beta titanium alloys, strength is developed during aging as a result of the precipitation of alpha phase in the beta matrix. Figure 9 depicts the as cold worked structure prior to aging, while in Fig. 10 the resultant microstructure after aging is shown. Aging heavily cold worked material results in uniform alpha precipitation through the thickness of the material. The observed decrease in strength as the aging temperature and time increases is a result of coarsening of the alpha precipitates.

TABLE 3—*Mechanical properties of Ti-15Mo round rod in various conditions.*

Temp. (°C)	Aging Time (h)	Cold Work (%)	Avg. U.T.S. (MPa)	Avg. Y.S. (MPa)	Avg. Elong. (%)	Avg. R.O.A. (%)	Ave. Mod. (GPa)
...[4]	...	...	762	558	46.4	73.9	78
...[5]	...	40 %	1154	672	18.0	75.6	57
	1	40 %	1560	1509	4.7	14.0	115
	2	40 %	1522	1433	6.5	9.3	114
	4	40 %	1473	1391	6.5	16.0	108
454[6]	8	40 %	1408	1300	7.3	25.5	107
	12	40 %	1365	1258	8.0	24.5	110
	16	40 %	1339	1217	8.8	26.0	108
	24	40 %	1303	1207	10.3	33.0	109
	1	40 %	1405	1272	9.0	27.7	106
	2	40 %	1366	1254	10.0	35.4	110
	4	40 %	1324	1234	11.0	30.0	98
482[6]	8	40 %	1300	1210	10.0	21.6	104
	12	40 %	1267	1190	12.0	31.5	102
	16	40 %	1259	1188	10.0	24.8	103
	24	40 %	1214	1140	12.0	30.8	94
	1	40 %	1315	1234	9.0	25.8	104
	2	40 %	1283	1206	10.0	26.2	101
	4	40 %	1233	1161	11.0	33.8	99
510[6]	8	40 %	1172	1096	12.0	45.5	111
	12	40 %	1174	1096	12.0	28.8	96
	16	40 %	1146	1060	15.0	39.0	104
	24	40 %	1101	1016	14.0	39.4	104
	1	40 %	1205	1130	11.3	34.5	100
	2	40 %	1177	1099	9.5	23.5	103
	4	40 %	1146	1061	12.0	22.5	102
538[6]	8	40 %	1061	986	15.0	42.5	105
	12	40 %	1053	971	13.5	37.0	103
	16	40 %	1018	933	15.5	48.5	99
	24	40 %	1001	915	16.0	42.5	103

[4] Mechanical properties of solution treated Ti-15Mo material.
[5] Mechanical properties of Ti-15Mo drawn with 40 % reduction of area.
[6] Mechanical properties of Ti-15Mo drawn with 40 % reduction of area and aged.

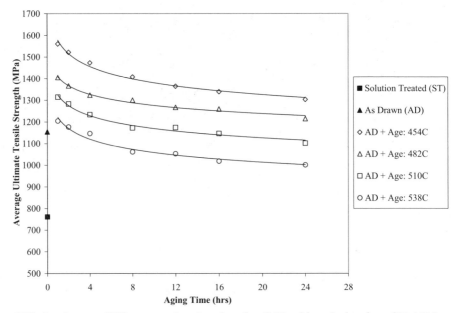

FIG. 7—*Average UTS versus aging time data for 40 % cold worked and aged Ti-15Mo.*

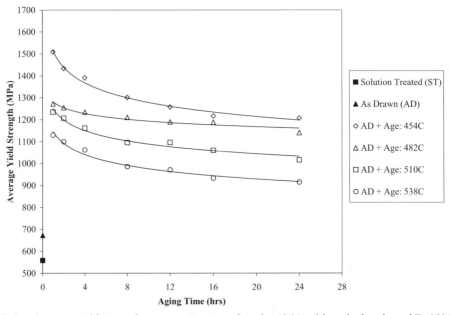

FIG. 8—*Average yield strength versus aging time data for 40 % cold worked and aged Ti-15Mo.*

FIG. 9—*Ti-15Mo as-drawn microstructure (40 %). Grain size is ASTM 5.5 (100×).*

FIG. 10—*Ti-15Mo that has been 40 % cold drawn and aged reveals uniform alpha precipitation within the microstructure (500×).*

The ductility results from the aging experiments are shown in Figs. 11 and 12. In Fig. 11, the percent elongation is shown as being greater than 9 % for all of the aging time and temperature combinations except when aging was performed at 454°C for less than 24 h. The reduction of area results shown in Fig. 12 depict greater than 20 % reduction of area for all time and temperature combinations except when aging at 454°C for less than 8 h. Additionally, although the data do reveal some overlap, the ductility increases as the aging temperature increases, as expected, due to the dependence of strength on aging temperature as discussed above.

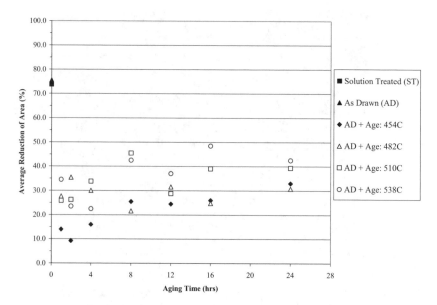

FIG. 11—*Average elongation versus aging time data for 40 % cold worked and aged Ti-15Mo.*

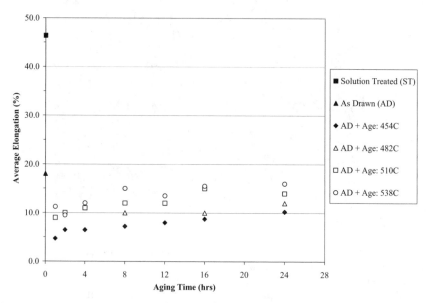

FIG. 12—*Average reduction of area versus aging time data for 40 % cold worked and aged Ti-15Mo.*

The dependence of modulus of elasticity on the aging time and temperature is depicted in Fig. 13. Figure 13 clearly shows that the modulus of elasticity slightly decreases with aging time. Additionally, it was found that aging at 454°C results in a higher modulus (stiffer) material than when aging was performed at temperatures of 482°C and higher.

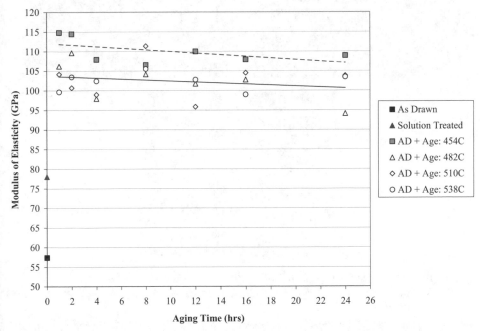

FIG. 13—*Modulus of elasticity versus aging time data for 40 % cold worked and aged Ti-15Mo. Modulus of elasticity of solution treated and as-drawn Ti-15Mo is shown for comparison.*

The fracture surfaces of room temperature tensile specimens comparing a solution treated sample to a high strength cold drawn and aged sample are depicted in Fig. 14. The fracture surface for material in the solution treated condition (Fig. 14a), as expected, revealed a classic cup and cone ductile fracture appearance where damage accumulation occurred by global void growth and coalescence prior to failure. Ductile fracture was also the dominant mode of failure for material that was cold drawn and aged, which was expected due to the high ductility values reported earlier. However, isolated pockets of cleavage fracture surrounded by a ductile cup and cone matrix were observed on the fracture surfaces of the aged material (see Fig. 14c). The isolated cleavage fracture was observed for samples treated to the aging cycle that produced the highest strength properties (454°C for 1 h) as well as the lowest strength properties (538°C for 24 h). It is important to note, however, that failure occurred predominantly by ductile fracture, and by no means is the material in the high strength condition considered brittle.

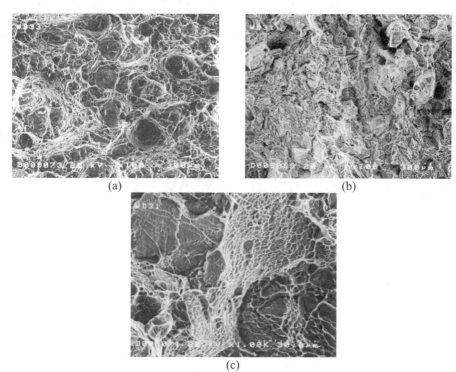

(a)   (b)

(c)

FIG. 14—(a) *Tensile fracture surface of solution treated Ti-15Mo.*
(b) *Tensile fracture surface of Ti-15Mo aged at 454°C for 1 h, i.e., highest strength condition.*
(c) *Small isolated areas of cleavage failure surrounded by ductile fracture in high strength Ti-15Mo.*

Conclusions

- In as-drawn material, the ultimate tensile strength increases with increasing cold reduction levels. Post age hardening, increasing the level of cold reduction has no influence on the resultant ultimate tensile strength.
- In as-drawn material, increasing cold reduction does not increase the yield strength above what is measured for solution treated properties. Post age hardening, increasing the level of cold reduction has no influence on the resultant yield strength.
- Elongation decreases, but reduction of area remains constant as cold reduction increases in as-drawn material. Post aging, both elongation and reduction of area increase with increasing cold work.
- Aging studies performed using 40 % cold reduced Ti-15Mo rod revealed that as the aging temperature increases, both the UTS and YS decrease.
- An extremely fast aging response was observed in this material where the highest strength is observed at the shortest aging time.

- As the hold time increases from 1 to 24 h during aging, the strength of the material decreases.
- As the aging temperature increases, the strength decreases and the ductility increases.
- The modulus of elasticity slightly decreases for as-drawn and aged material as the aging time increases.
- Aging at 454°C results in a higher modulus (stiffer) material than when aging was performed at temperatures of 482°C and higher.
- Isolated pockets of cleavage fracture surrounded by a ductile cup and cone matrix were observed on the fracture surfaces of all cold drawn and aged high strength material. Nevertheless, tensile fracture in high strength Ti-15Mo is predominantly ductile.

References

[1] ATI Allvac® Ti-15Mo Beta Titanium Alloy, Technical Data Sheet, ATI Allvac, Monroe, NC, 16 June 2004 (http://www.allvac.com/allvac/pages/Titanium/Ti15Mo.htm).

[2] Zardiackas, L. D., Mitchell, D. W., and Disegi, J. A., "Characterization of Ti-15Mo Beta Titanium Alloy for Orthopaedic Implant Applications," *ASTM STP 1272, Medical Applications of Titanium and Its Alloys: The Material and Biological Issues*, S. A. Brown and J. E. Lemons, Eds., ASTM International, West Conshohocken, PA, 1996, pp. 60–75.

[3] Bogan, J., Zardiackas, L., and Disegi, J., "Stress Corrosion Cracking Resistance of Titanium Implant Materials," *Transactions 27th Annual Meeting of the Society for Biomaterials*, St. Paul, MN, April 24–29, 2001.

[4] Disegi, J. and Zardiackas, L., "Metallurgical Features of Ti-15Mo Beta Titanium Alloy for Orthopaedic Trauma Applications," *Medical Device Materials, Proceedings from the Materials & Processes for Medical Devices Conference*, 8–10 September, 2003, ASM International, Materials Park, OH, 2004, pp. 337–342.

[5] ASTM Standard F 2066, "Standard Specification for Wrought Titanium-15Molybdenum Alloy for Surgical Implant Applications (UNS R58150)," ASTM International, West Conshohocken, PA.

[6] ASTM Standard F 1472, "Standard Specification for Wrought Titanium-6Aluminum-4Vanadium Alloy for Surgical Implant Applications (UNS R56400)," ASTM International, West Conshohocken, PA.

[7] ASTM Standard F 136, "Standard Specification for Wrought Titanium-6Aluminum-4Vanadium ELI (Extra Low Interstitial) Alloy for Surgical Implant Applications (UNS R56401)," ASTM International, West Conshohocken, PA.

[8] ASTM Standard F 1295, "Standard Specification for Wrought Titanium-6Aluminum-7Niobium Alloy for Surgical Implant Applications [UNS R56700]," ASTM International, West Conshohocken, PA.

[9] ASTM Standard F 1713, "Standard Specification for Wrought Titanium-13Niobium-13Zirconium Alloy for Surgical Implant Applications (UNS R58130)," ASTM International, West Conshohocken, PA.

[10] ASTM Standard F 1813, "Standard Specification for Wrought Titanium-12Molybdenum-6Zirconium-2Iron Alloy for Surgical Implant Applications (UNS R58120)," ASTM International, West Conshohocken, PA.

[11] Allvac Titanium 6Al-4V Alloy, Technical Data Sheet, ATI Allvac, Monroe, NC, 7 June 2004.

[12] Mishra, A., Davidson, J., Poggie, R., Kovacs, P., and FitzGerald, T., "Mechanical and Tribological Properties and Biocompatibility of Diffusion Hardened Ti-13Nb-13Zr - A New Titanium Alloy for Surgical Implants," *ASTM STP 1272, Medical Applications of Titanium and Its Alloys: The Material and Biological Issues*, S. A. Brown and J. E. Lemons, Eds.,

ASTM International, West Conshohocken, PA, 1996, pp. 96–113.

[13] Wang, K., Gustavson, L., and Dumbleton, J., "Microstructure and Properties of a New Beta Titanium Alloy, Ti-12Mo-6Zr-2Fe, Developed for Surgical Implants," *ASTM STP 1272, Medical Applications of Titanium and Its Alloys: The Material and Biological Issues*, S. A. Brown and J. E. Lemons, Eds., ASTM International, West Conshohocken, PA, 1996, pp. 76–87.

[14] ASTM Standard F 1798, "Standard Guide for Evaluating the Static and Fatigue Properties of Interconnection Mechanisms and Subassemblies Used in Spinal Arthrodesis Implants," ASTM International, West Conshohocken, PA.

[15] ASTM Standard F 2193, "Standard Specifications and Test Methods for Components Used in the Surgical Fixation of the Spinal Skeletal System," ASTM International, West Conshohocken, PA.

[16] ASTM Standard F 1717, "Standard Test Methods for Spinal Implant Constructs in a Vertebrectomy Model," ASTM International, West Conshohocken, PA.

[17] Shamblen, C. E. and Hunter, G. B., *Titanium Base Alloys - Clean Melt Process Development*, Iron and Steel Society, Inc., L. H. Lherbier and J. T. Cordy, Eds., Warrendale, PA, 1989, pp. 3–11.

[18] Davis, R. M. and Forbes Jones, R. M., "Manufacturing Processes for Semi-Finished Titanium Biomedical Alloys," *ASTM STP 1272, Medical Applications of Titanium and Its Alloys: The Material and Biological Issues*, S. A. Brown and J. E. Lemons, Eds., ASTM International, West Conshohocken, PA, 1996, pp. 17–29.

[19] Marquardt, B., "Development and Characterization of High-Strength Allvac Ti-15Mo Alloy Bar and Rod for Spinal Innovations Inc.," ATI Allvac Internal Report, Jan. 2002.

ALLOY PROPERTIES

Journal of ASTM International, June 2005, Vol. 2, No. 6
Paper ID JAI12781
Available online at www.astm.org

Marie Koike,[1] *Qing Guo,*[2] *Mikhail Brezner,*[3] *Hideki Fujii,*[4] *and Toru Okabe*[5]

Mechanical Properties of Cast Ti-Fe-O-N Alloys

ABSTRACT: A recent addition to existing titanium alloys is the Super-TIX™ series (Nippon Steel Corp., Japan), which has intermediate strength between the strengths of CP Ti and Ti-6Al-4V as a result of balancing the Fe, O, and N concentrations. Two alloys in this series were investigated: Ti-1%Fe-0.35%O-0.01%N (Super-TIX800™; LN) and Ti-1%Fe-0.3%O-0.04%N (Super-TIX800N™; HN). These alloys were arc-melted and cast using an investment casting method, and their mechanical properties were examined. The yield strength (~600 MPa) and tensile strength (~680 MPa) were approximately 33 and 29 % higher, respectively, than the corresponding strength of cast CP Ti (ASTM Grade 2). Their percent elongation was somewhat (but not significantly) higher (2–3 %) than that of Ti-6A-4V. On the other hand, the elongation of CP Ti was approximately 7 %. The moduli of elasticity of the alloys ranged from 100–120 GPa. These Ti-Fe-O-N alloys exhibit higher uniaxial yield and tensile strengths than CP Ti, which is currently used for dental applications.

KEYWORDS: titanium, titanium alloys, mechanical properties, biocompatibility

Introduction

Good mechanical properties, plus excellent biocompatibility and corrosion resistance, have made titanium a very effective material for biomedical applications. However, when the application requires better mechanical properties than those offered by commercially pure (CP) titanium, stronger alloys, such as Ti-6Al-4V or Ti-6Al-7Nb, may be preferred. A series of titanium alloys (Super-TIX™) with an intermediate range of strength between that of CP Ti and Ti-6Al-4V was recently developed by Nippon Steel Corp., Japan [1]. The wrought form of these alloys has a wide range of tensile strengths (700–1000 MPa) with 10–30 % elongation, depending on their chemical compositions, working conditions (cold or hot), and heat treatment. Two of these alloys include Super-TIX800™ (LN) (Ti-1%Fe-0.35%O-0.01%N) and Super-TIX800N™ (HN) (Ti-1%Fe-0.3%O-0.04%N). These are two-phase $\alpha+\beta$ alloys since they contain Fe as a β stabilizer and O and N as α stabilizers [2]. The particular alloy properties were obtained by balancing these three elements.

Since the Ti-Fe-O-N alloys do not contain expensive alloying elements, they are reportedly cheaper to manufacture [1]. In addition, these alloys are acceptable for biomedical applications,

Manuscript received 12 August 2004; accepted for publication 24 November 2004; published June 2005. Presented at ASTM Symposium on Titanium, Niobium, Zirconium, and Tantalum for Medical and Surgical Applications on 9-10 November 2004 in Washington, DC.
[1] Assistant Professor, Nagasaki University Graduate School of Biomedical Sciences, Nagasaki, Japan.
[2] Research Assistant I, Baylor College of Dentistry, Texas A&M University System Health Science Center, Dallas, TX 75246.
[3] Research Assistant I, Baylor College of Dentistry, Texas A&M University System Health Science Center, Dallas, TX 75246.
[4] Chief Researcher, Steel Research Labs, Nippon Steel Corp., Chiba, Japan.
[5] Regents Professor and Chairman, Baylor College of Dentistry, Texas A&M University System Health Science Center, Dallas, TX 75246.

103

since they do not contain any elements that are questionable in terms of harmful biological effects [3]. Originally, they were intended for use in the wrought form. However, we thought it worthwhile to characterize various properties of these alloys in their cast form to determine their suitability for dental applications. In the present study, the mechanical properties of the cast Ti-Fe-O-N alloys were evaluated and compared to those of Ti-6Al-4V and CP Ti, which have a long history of biomedical use.

Materials and Methods

Metals Used and Casting Procedure

The titanium alloys studied were Super-TIX800™ (LN) and Super-TIX800N™ (HN) (both manufactured by Nippon Steel Corp., Chiba, Japan). Commercially pure titanium (CP Ti: ASTM Grade 2, Titanium Industries, Grand Prairie, TX) and Ti-6Al-4V (ASTM Grade 5, Titanium Industries) were used as controls. The chemical compositions of all metals tested are listed in Table 1.

TABLE 1—*Chemical compositions of alloys tested.*

Alloy	Nominal Compositions	Analyzed Compositions (mass %)
Super-TIX800™ (LN)	Ti-1%Fe-0.35%O-0.005%N	0.91Fe; 0.37O; 0.005N; 0.004C; 0.006Si; 0.013Ni; 0.006Sn; 0.007Al; 0.012Cr; balance: Ti
Super-TIX800N™ (HN)	Ti-1%Fe-0.3%O-0.04%N	0.96Fe; 0.30O; 0.041N; 0.004C; 0.009Si; 0.013Ni; 0.007Sn; 0.008Al; 0.012Cr; balance: Ti
CP Ti ASTM Grade 2		Fe < 0.0800; O < 0.1400; N < 0.0060; C < 0.0090; H < 0.0006; balance Ti
Ti-6Al-4V Grade 5		Fe < 0.1800; O < 0.180; N < 0.0100; C < 0.0100; H < 0.0016; Al < 6.100; V < 3.970; balance: Ti

Discs (30 mm diameter, 10 mm thick) of each metal (approximately 30 g) were used to cast the specimens. Each alloy was arc-melted and cast into a magnesia-based mold (Selevest CB, Selec, Osaka, Japan) at 200°C in a centrifugal casting machine (Ticast Super R, Selec). The main constituents of the investment material were MgO, CaO, and Al_2O_3, plus Zr powder. The casting chamber was first evacuated to 7×10^{-2} Pa (5×10^{-4} Torr) and then filled with high-purity argon gas (99.999 %, Ultra High Purity Argon, Gasco, Dallas, TX) until the pressure reached 2.8×10^4 Pa (200 Torr). An electric current of 200A was used to melt LN, HN, and CP Ti, and 190A was used for Ti-6Al-4V. The melting time was 55 s.

Two types of wax patterns were invested to make the castings: a wax slab for the hardness test, metallurgical examination, and X-ray diffractometry (10 mm × 10 mm × 2.5 mm) and a dumbbell-shaped plastic pattern for testing mechanical properties (20 mm gauge length; 2.8 mm diameter). The internal porosity of the cast specimens was examined with a conventional dental X-ray unit (Dentsply Gendex GX900, Des Plaines, IL) under the following conditions: target film distance, 50 cm; tube voltage, 70 kVp; tube current, 15 mA; exposure, 1–2 s. Specimens with noticeably large pores were excluded from further testing. The minimum diameter of pores detected by this method was estimated to be 0.05 mm.

Metallography and Fractography

A minimum of two pieces of each cast metal was metallurgically polished in a slurry of suspended 0.05 μm colloidal silica (Colloidal Silica Suspension, Allied High Tech Products, Inc., Rancho Dominguez, CA) after polishing to #2000 abrasive paper (Allied High Tech Products, Inc.). The finished specimens were etched using a hydrofluoric acid-based solution (Keller's reagent: 2 ml HF, 3 ml HCl, 5 ml HNO_3, 190 ml H_2O)[4]. The prepared surfaces were examined using an optical microscope (Epiphot 200, Nikon, Japan) and a scanning electron microscope (JSM-6300, JEOL, Japan) (SEM). A Si (Li) X-ray detector (Noran Instruments, Middleton, WI) and an X-ray microanalysis and digital image system (5480; IXRF Systems, Houston, TX) controlled by a work station (EDS 2000; IXRF Systems) were used for the energy dispersive spectrometer (EDS) analysis. In addition, the fractured surfaces of a minimum of two specimens for all metals tested was examined using SEM after tensile testing.

X-Ray Diffractometry

For each cast specimen of each metal, one of the square cast surfaces (10 mm × 10 mm) of each specimen was polished to a depth of approximately 300 μm using up to #2000 abrasive paper to ensure that the diffractometry was performed in the interior structure. X-ray diffractometry (XRD) (Miniflex 2005 X-ray Diffractometer, Rigaku, Tokyo, Japan) was carried out using Cu Kα radiation (Ni-filtered) generated at 30 kV and 10 mA and a scanning rate of 0.2° per min. At least two specimens were examined for each metal. The peaks of the XRD patterns were indexed to the ICDD (International Center for Diffraction Data, Swarthmore, PA) polycrystalline powder diffraction files [5].

Tensile Testing

Tensile testing of all the metals was conducted at room temperature at a strain rate of $170\mu s^{-1}$ (crosshead speed of 0.01 in./min) on a universal testing instrument (Model No. 1125, Instron, Canton, MA). Four specimens were tested for each metal. Ultimate tensile strength, yield strength at 0.2 % offset, modulus of elasticity, and elongation to failure were determined. The results were analyzed using one-way ANOVA and the Scheffé's test at $\alpha = 0.05$.

Microhardness Measurements

Changes in Vickers hardness with depth were evaluated for the polished cross-sections of each cast alloy using a microhardness tester (FM-7, Future Tech, Tokyo, Japan) with a 200 g load and 15 s dwell time starting at 25 μm below the cast surface, then at 50 μm, and every 50 μm thereafter to a depth of 500 μm. Two specimens of each alloy were used to measure the hardness. To obtain the hardness numbers at each depth, four measurements were made at randomly chosen locations for each alloy. Thus, there were eight measurements in all at each depth. The bulk hardness values at 300 μm were analyzed using one-way ANOVA and the Scheffé's test ($\alpha = 0.05$).

Results

X-Ray Diffraction

Figure 1 summarizes typical X-ray diffraction patterns for LN, HN, CP Ti, and Ti-6Al-4V. All the major peaks were indexed as α Ti (card file #44-1294). Although the β Ti peaks (card file #44-1288) were not as prominent as some of the α Ti peaks, the major β Ti peaks in Ti-6Al-4V, LN, and HN could be identified. These were from diffraction planes such as (111) (2θ, 38.48°), (200) (2θ, 55.54°), and (211) (2θ, 69.61°).

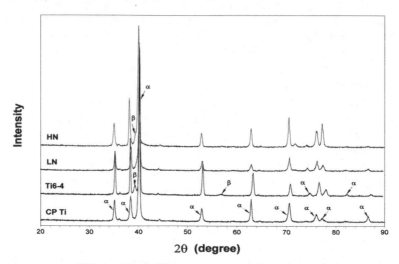

FIG. 1—*X-ray diffraction patterns of metals tested.*

Microstructures

Typical SEM microstructures of cast LN, HN, CP Ti, and Ti-6Al-4V specimens are shown in Figs. 2, 3, 4, and 5, respectively. Figures 2a–5a show the microstructures near the cast surface for each metal, and views of the interior are seen in Figs. 2b–5b. The layer near the cast surface in all the metals consists of inwardly grown α grains (dark areas). The interior structures of all the metals have a Widmanstätten basketweave appearance. In all alloys, the platelets (dark areas) are α phase, and the light areas between these plates are β phase. The EDS surveys of the LN and HN specimens showed Fe in the area that we presumed was the β phase. No Fe was detected in the α phase areas. Both CP Ti and Ti-6Al-4V had Widmanstätten structures with the α and β phases. CP Ti should contain mostly α phase, but there may be some β phase between the α plates [6].

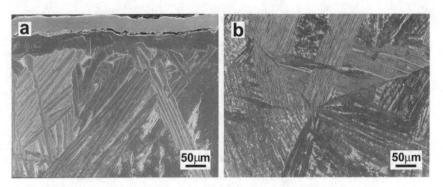

FIG. 2—*Microstructures of LN:* a) *near the cast surface,* b) *interior.*

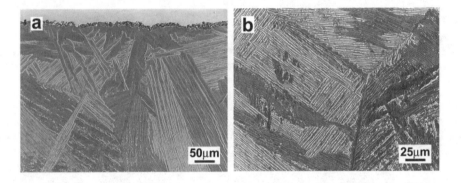

FIG. 3—*Microstructures of HN:* a) *near the cast surface,* b) *interior.*

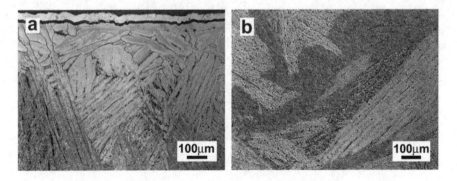

FIG. 4—*Microstructures of CP-Ti:* a) *near the cast surface,* b) *interior.*

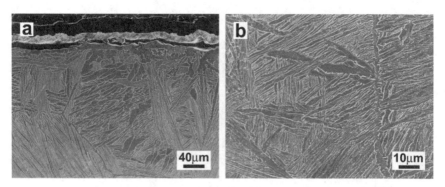

FIG. 5—*Microstructures of Ti-6Al-4V:* a) *near the cast surface,* b) *interior.*

Fractured Surfaces

Three different views of the fractured LN and CP Ti specimens are shown as examples in Figs. 6a–6c (LN) and Figs. 7a–7c (CP Ti): the exterior near the fractured areas (Figs. 6a and 7a), the near-surface area of the fractured surfaces (Figs. 6b and 7b), and the bulk area of the fractured surfaces (Figs. 6c and 7c). Figure 6a shows some transverse cracks that developed in LN during tensile testing; similar cracks were found in the HN and Ti-6Al-4V specimens. However, there were more cracks in the CP Ti specimen (Fig. 7a), which was more ductile. In Figs. 6b and 7b, a brittle, transgranular cleavage fracture mode dominated the surface of the castings. On the other hand, a ductile fracture mode with microvoid coalescence was prevalent in the interior of the castings (Figs. 6c and 7c). Although the micrographs of the fractured HN and Ti-6Al-4V specimens are not shown, the fracture mode was similar to that found for the LN and CP Ti specimens.

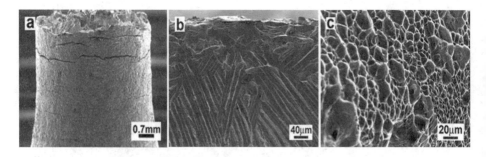

FIG. 6—*SEM micrographs of LN:* a) *exterior near the fractured area,* b) *near the surface of fractured area,* c) *bulk area.*

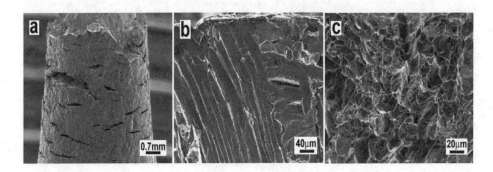

FIG. 7—*SEM micrographs of CP Ti;* a) *exterior near the fractured area;* b) *near the surface of fractured area;* c) *bulk area.*

Mechanical Properties

The yield strength, tensile strength, and elongation for the metals tested are summarized in Fig. 8. As indicated, the yield strength and tensile strength of LN and HN were almost identical. The strengths of both LN and HN were significantly greater than those of CP Ti and less than those of Ti-6Al-4V. As for the elongation, the CP Ti had significantly higher values than did the rest of the metals. There were no significant differences in the elongation among LN, HN, and Ti-6Al-4V. Although not included in Fig. 8, the moduli of elasticity of all metals tested were not statistically different; the values ranged from 100–120 GPa.

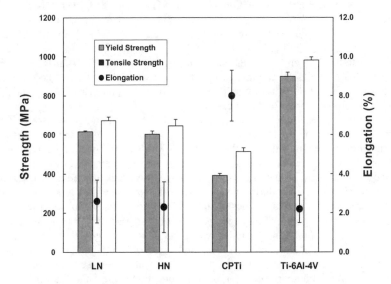

FIG. 8—*Mechanical properties results of metals tested.*

Figure 9 displays the changes in microhardness with depth from the cast surface for the metals tested. The interior bulk hardness (the mean of microhardness at the depths of 300, 400, and 500 μm below the cast surface) of CP and Ti-6Al-4V (Ti-64) were statistically the lowest and highest, respectively, among the metals tested. The greater hardness near the cast surfaces indicated that the hardened surface layer (α-case) formed on all of the metals.

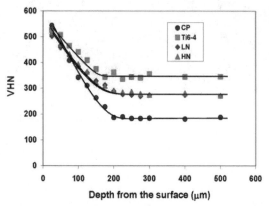

FIG. 9—*Microhardness of metals tested.*

Discussion

LN and HN are available for industrial use in the form of wrought rolled rods, sheets, and pipes. In the annealed form after cold- or hot-rolling, LN had yield and tensile strengths of 600 and 700 MPa, respectively, and elongation of 20–30 % [1]. In general, the mechanical properties of as-cast metals are inferior to those of the corresponding wrought metals. However, in the present study, the yield and tensile strengths of both the as-cast LN and HN alloys were almost the same as for the wrought metal (600 and 700 MPa, respectively) (Fig. 8). These strengths were approximately 33 % and 29 % higher than for the as-cast CP Ti. On the other hand, the strength of the cast Ti-6Al-4V was much higher than that of LN and HN. The strength of the wrought form of LN and HN was designed to fall between the strengths of CP Ti and Ti-6Al-4V [1]; the same results were true for the as-cast alloys.

As in most cases, the increase in strength resulted in reduced elongation, which was a little more than 2 % for both the LN and HN alloys. This amount was similar to that of the as-cast Ti-6Al-4V (in contrast, the elongation for the cast CP Ti was 8 %). It should be noted that the SEM observations of the fractured surfaces after tensile testing (for example, in LN as shown in Fig. 6c) showed evidence of considerable ductile fracture. The premature transverse cracks on the sides of the specimens created during tensile testing (for example, in LN as shown in Fig. 6a) reduced the strength and ductility. These cracks developed by promoting microcrack initiation during tensile testing because of an incompatible brittle surface layer (α-case) and a ductile interior matrix. In our preliminary studies, we found that the oxygen concentration in the cast specimens of both LN and HN increased compared to the concentration before casting. Thus, it is likely that the concentration of oxygen in the interior also increased after casting. Oxygen contamination could be another reason for the low elongation values of the cast metals.

The trend for the bulk hardness of these alloys was similar to that found for the strength (Fig. 8). The microhardness profiles (Fig. 9) determined from the cast surfaces to the interior structures indicated the presence of the typical hardened surface layer (α-case), which mainly resulted from the reaction at the interface of the investment material and the solidifying metal. In most cases, our EDS examination showed Al and Zr, which appeared to diffuse into the cast structure after the decomposition of these oxides by the molten Ti. This layer appeared to be 200 μm thick on LN, HN, and CP Ti, whereas in the Ti-6Al-4V, it was approximately 150 μm. A thinner α-case is typical for alloys with a considerable amount of alloying elements; the thermodynamic explanation for this phenomenon can be found in Takada et al. [7].

The increased strength of both LN and HN compared to CP Ti is due to alloying Fe, O, and N, which resulted in the two-phase α+β microstructure. Iron is a strong β stabilizing element in titanium [2] and effectively increases the strength through solid-solution hardening [8], as do O and N, which are α stabilizers [2]. It is well known that the O and Fe contents determine the level of strength of CP Ti [2]. The microstructures of both LN and HN showed the formation of the Widmanstätten structure consisting of acicular α plates formed at and growing from prior β boundaries (Figs. 2b and 3b). We believe the areas between the α plates consist of β phase. The minimum amount of Fe necessary to retain the β phase is reported to be 3.5 % [9]; however, the 1 % Fe in the present alloys was apparently high enough to retain some β phase in the as-cast structure. Nishiyama et al. [10] reported that Ti-3%Fe formed a martensite structure in the quenched structure. However, a diffusional transformation apparently occurred in the present alloys in the cast and bench-cooled condition (still in the mold), although some diffusionless phase transformation may have taken place. The formation of the omega phase in the present specimens is unlikely, although Guseva and Dolinskaya [11] found the omega phase in a quenched Ti-Fe alloy containing more than 2.3 % Fe.

Because of the abundant solubility of the present alloying elements (Fe, O, N) in the β phase [12], all of these elements are dissolved in the β Ti lattice at an early stage of cooling above the β transus. However, the situation changes below the β transus because of an abrupt reduction in the solubility of Fe in the α titanium (0.047 % at the eutectoid temperature of 595°C) [12]. Thus, the formation of the α plates in the β grains leads to the Fe enrichment in the retained β phase. The diffusion of Fe in the β lattice is one of the fastest among the titanium alloys containing transition metals [13]. This phenomenon also helps all of the Fe to stay in the β phase region. As described above, we found Fe in the β phase area in both LN and HN. No Fe was found in the α phase area. The improved strength of these alloys comes first of all from the refined two-phase α+β microstructure. In addition, the interstitial solid-solution strengthening of the α phase grains by O and N, and the substitutional solid-solution strengthening of the β phase grains by Fe, further contribute to the reaction. The similar strength of LN and HN indicated that strengthening with the increased N content balanced the decreased O content. It was found that increased N in the wrought form of HN increased the corrosion resistance when tested in a boiled NaCl+HCl solution [14].

Summary

Two Ti-Fe-O-N alloys were cast using a titanium casting machine used in dentistry. The yield and tensile strengths of these alloys were approximately 33 and 29 % higher, respectively, than for CP Ti (ASTM Grade 2), which is currently used for fabricating various types of dental prostheses.

Acknowledgments

The authors wish to thank Mrs. Jeanne Santa Cruz for editing the manuscript. This work was partially supported by NIH/NIDCR grant 11787.

References

[1] Fujii, H., Soeda, S., Hanaki, M., and Okano, H., "Development of High-Performance, Low Alloy Ti-Fe-O-N Series," P. A. Blenkinsop, W. J. Evans, H. M. Flower, Eds., *Titanium '95, Science and Technology*, Cambridge: The Institute of Metals, 1996, pp. 2309–2316.
[2] Donachie, M. J., *Titanium: A Technical Guide*, 2nd ed., ASM, Materials Park, OH, 2000.
[3] Fanning, J. C., "Properties and Processing of a New Metastable Beta Titanium Alloy for Surgical Implant Applications: Timetal 21SRx," P. A. Blenkinsop, W. J. Evans, H. M. Flower, Eds., *Titanium '95, Science and Technology*, Cambridge: The Institute of Metals, 1996, 1800–1807.
[4] ASM Handbook Committee, *Metals Handbook*, 8th ed., Vol. 7., Atlas of Microstructures of Industrial Alloys, American Society for Metals, Metals Park, OH, 1972.
[5] International Center for Diffraction Data, Powder diffraction file, set 1-48, JCPDS International Center for Diffraction Data, Swarthmore, PA, 1998.
[6] Boyer, R., Welsch G., and Collings E.W., Eds., *Materials Properties Handbook, Titanium Alloys*, Materials Park, OH: ASM International, 1994, p. 176.
[7] Takada, Y., Nakajima, H., Okuno, O., Okabe, T., "Microstructure and Corrosion Behavior of Binary Titanium Alloys with Beta-Stabilizing Elements," *Dent Mater J*, 20:34–52, 2001.
[8] Okuno, O., Shimizu, A., and Miura, I., "Fundamental Study on Titanium Alloys for Dental Casting," *Journal of the Japanese Society for Dental Materials and Devices*, 4:708–715, 1985.
[9] Bania, P. J., "Beta Titanium Alloys and Their Role in the Titanium Industry," Eylon, D., Boyer, R.R., Koss, D.A., *Beta Titanium Alloys in the 1990s*, Warrendale, PA: The Minerals, Metals and Materials Society, 1993, pp. 3–14.
[10] Nishiyama, Z., Oka, M., and Nakagawa H., "Transmission Electron Microscope Study of the Martensites in a Titanium-3 wt% Iron Alloy," *J Japan Inst of Metals*, 30:16–21, 1966.
[11] Guseva, L. N. and Dolinskaya, L. K., "Metastable Phases in Titanium Alloys with Group VIII Elements Quenched from the β Region," *Russian Metallurgy*, 6:155–159, 1974.
[12] Murray, J. L., Ed., *Phase Diagrams of Binary Titanium Alloys*, ASM International, Metals Park, OH, 1987.
[13] Collings, E. W., *The Physical Metallurgy of Titanium Alloys*, ASM, Metals Park, OH, 1984.
[14] Fujii, H., Yamashita, Y., Hatta, Y., and Matsuhashi K., "Materials Properties and Applications of Super-TIX High-Performance Titanium Alloy Series," *Titanium*, 49:171–176, 2001 (in Japanese).

Journal of ASTM International, July/August 2005, Vol. 2, No. 7
Paper ID JAI12782
Available online at www.astm.org

Ikuya Watanabe,[1] *Jie Liu,*[2] *Akihide Saimoto,*[3] *Jason Griggs,*[4] *and Toru Okabe*[5]

Effect of Surface Reaction Layer on Three-Point Flexure Bond Strength of Resin Composite to Cast Ti and Ti-6Al-7Nb

ABSTRACT: This study investigated the effect of the α-case on the three-point flexure bond strength of resin composite to different thicknesses of cast CP Ti and Ti-6Al-7Nb. Cast plates (0.3, 0.5, and 0.7 mm thick) were prepared for each metal. Plates without α-case were also prepared for the 0.5 and 0.7 mm-thick CP Ti and Ti-6Al-7Nb. After resin composite was polymerized on the plates, the resin/metal specimens were subjected to three-point bending testing. The fracture force (N) and deflection (mm) at fracture were recorded. The fracture force of the specimens increased with increasing specimen thickness. Specimens without α-case withstood greater fracture force than specimens of either metal with α-case. CP Ti and Ti-6Al-7Nb without α-case deflected more than with α-case. The results suggested that the bonding between the veneer composite and thin cast structures (<0.5 mm) is likely to fail and that the surface reaction layer (α-case) should ideally be removed.

KEYWORDS: three-point bending test, resin composite, cast titanium, titanium alloy

Introduction

Cast pure titanium and its alloys have been used increasingly to make fixed and removable dental prostheses [1–4] due to the development of various materials and improved technologies for titanium casting. One of the problems associated with titanium casting is that the chemical reactivity of titanium with elements in conventional investments, particularly with silicate-based or phosphate-bonded investment [5], produces a hard, brittle reaction layer (α-case) on the cast surface [6,7]. Therefore, the refractory components of investment materials specifically made for titanium casting include oxides (such as calcia, magnesia, and zirconia [8–10]) that are more stable than titanium oxides to prevent a reaction layer from forming. Although these investments reduce the reaction layer and decrease surface roughness, surface contamination cannot be entirely avoided; thus, a reaction zone exists to some degree on each casting [8,10].

Resin composite-veneered adhesive fixed partial dentures (AFPD) with cast metal retainers (0.3–0.7 mm thick) are sometimes applied to anterior restorations. Although the reaction layers on the cast surface reportedly reduce the ductility and fatigue resistance of removable partial denture frameworks and clasps [11], they also increase the bending strengths of the denture plate frameworks and produce high resistance to bending stress in the flexible, thin retainers of AFPDs

Manuscript received 12 August 2004; accepted for publication 10 January 2005; published July 2005. Presented at ASTM Symposium on Titanium, Niobium, Zirconium, and Tantalum for Medical and Surgical Applications on 9-10 November 2004 in Washington, DC.
[1] Assistant Professor, Baylor College of Dentistry, Texas A&M University System Health Science Center, Dallas, TX 75246.
[2] Instructor, Nagasaki University Graduate School of Biomedical Sciences, Nagasaki, Japan.
[3] Associate Professor, Faculty of Engineering, Nagasaki University, Nagasaki, Japan.
[4] Associate Professor, Baylor College of Dentistry, Texas A&M University System Health Science Center, Dallas, TX 75246.
[5] Regents Professor and Chairman, Baylor College of Dentistry, Texas A&M University System Health Science Center, Dallas, TX 75246.

113

such as Maryland bridges [12] (Fig. 1). Strong resistance to bending stress is necessary for cast metal frameworks to withstand occlusal force and may also affect the bond strengths between metal frameworks and resin composite.

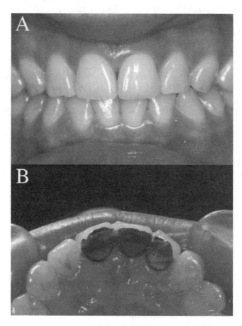

FIG. 1—*Resin composite-veneered adhesive fixed partial denture with thin metal framework retainers of the adjacent teeth for a left central incisor:* A) *frontal view;* B) *occlusal view.*

Generally, the bond strength of polymeric material to a metal substrate is measured by shear bond testing [13,14] or tensile bond testing [15,16] using various sample sizes. A three-point flexure bond test has been used [17–22] to investigate the bonding of ceramics to metal substrates. However, there is little information about the bond strength of dental resin composite polymer material retained to titanium and its alloys using three-point flexure bond testing.

The purpose of this study was to investigate the effect of the surface reaction layer on the three-point flexure bond strength of resin composite to cast Ti and Ti-6Al-7Nb, and to compare the bond strength with that to Co-Cr alloy, which is commonly used to fabricate metal partial denture frameworks.

Materials and Methods

The three metals used in this study [commercially pure (CP) Ti, Ti-6Al-7Nb alloy (T-Alloy Tough) and Co-Cr alloy (Cobaltan)] are listed in Table 1, as well as the specimen sizes. The thicknesses of 0.3, 0.5, or 0.7 mm correspond to the commonly used thicknesses for metal AFPD frameworks. Five types of plate patterns were prepared for the cast CP Ti and Ti-6Al-7Nb alloy. Figure 2 presents a diagram of the cast plate specimens.

TABLE 1—*Metals and specimen sizes employed in this study.*

	CP Ti	Ti-6Al-7Nb	Co-Cr
Metal	Ti grade 2 (Selec, Osaka, Japan)	T-Alloy Tough (GC, Tokyo, Japan)	Cobaltan (Shofu, Kyoto, Japan)
Composition (wt%)	Ti: 99.5; other: 0.5	Ti: 86.5; Al: 6.0; Nb:7.0; other: 0.5	Co: 63.0; Cr:29.0; Mo:6.0; other: 2.0
$d_M \times w$ (I_M = 25.0 mm)		0.3×5.0 0.5×5.0 $0.5\ (0.8\text{-}\alpha) \times 5.0\ (5.3\text{-}\alpha)$ 0.7×5.0 $0.7\ (1.0\text{-}\alpha) \times 5.0\ (5.3\text{-}\alpha)$	0.3×5.0 0.5×5.0 0.7×5.0

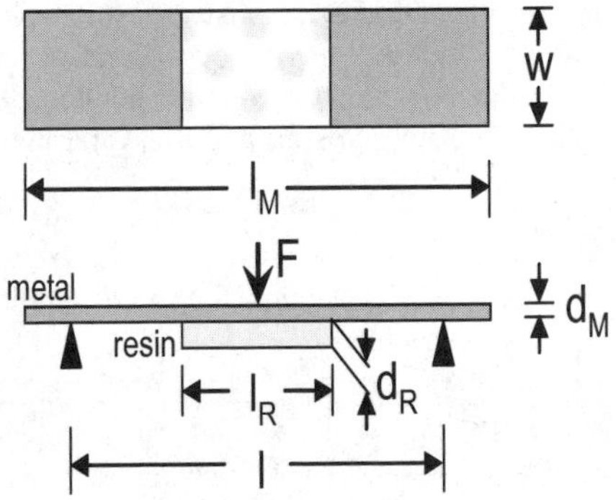

FIG. 2—*Diagram of the three-point flexural test specimen (dimensions of the test configuration are: l_M = 25 mm, l_R = 8 mm; l = 20 mm, d_M = 0.3, 0.5, 0.7 mm, d_R = 1 mm, w = 5 mm).*

The CP Ti and Ti-6Al-7Nb were cast using a centrifugal casting unit (Selecast System, Selec Co., Osaka, Japan) with a magnesia-based investment material (Selevest CB, Selec Co.). The Co-Cr alloy was cast with a phosphate-bonded investment (Snow White; Shofu, Kyoto, Japan) in a high-frequency casting machine (Argon Caster T; Shofu). The patterns for each metal were invested in mold rings. The overall investment burn-out schedule before casting followed the manufacturers' instructions. The mold temperatures for the titanium (CP Ti and alloy) and Co-Cr alloy at the time of casting were 200°C and 800°C, respectively. After casting, the molds of each metal were allowed to bench-cool to room temperature. The specimens were then retrieved from the investments and ultrasonically cleaned with distilled water for 10 min.

All of the cast CP Ti and Ti-6Al-7Nb plates were examined for internal porosity using a dental x-ray unit (Coronis 90, Asahi Roentgen Ind. Co., Ltd., Kyoto, Japan) before performing further testing. Any casting having internal porosity was eliminated from the experiments. The CP Ti and Ti-6Al-7Nb plates with dimensions of 0.8 (d_M) × 5.3 (w) × 25.0 (l_M) mm and 1.0 × 5.3 × 25.0 mm were abraded with No. 60 to No. 600 wet SiC sandpaper to create specimens without the α-case; 150 μm (0.15 mm) was removed from all surfaces except for both ends of the plate.

These cast plates were finished to the dimensions of 0.5 (d_M) × 5.0 (w) × 25.0 (l_M) mm and 0.7 × 5.0 × 25.0 mm, respectively.

After the surfaces to be bonded were sandblasted with 50 μm Al_2O_3 particles for 30 s at an air pressure of 5 kgf/cm², opaque resin (Dentacolor, Heraeus Kulzer, Wehrheim, Germany) was applied to an area [8 (l_R) × 5.0 (w) mm] in the center of the metal plate and light-polymerized in a xenostroboscopic unit (Uni XS, Heraeus Kulzer) for 180 s. The thickness of the opaque resin was approximately 200 μm (two layers, 90 s polymerization for each layer). The dentin resin (approximately 1 mm thick) was then built up using a jig and polymerized for 60 s. After polymerization, the surface of the dentin resin was polished with SiC paper and finally finished to a thickness (d_R) of 1 mm.

The resin/metal specimens were then subjected to a three-point bending test, which was conducted using a universal testing machine (AGS-10kNG, Shimadzu Corp., Kyoto, Japan) at a crosshead speed of 1.5 mm/min and a distance (l) of 20 mm between the support rods (Fig. 2). The specimens were loaded at the center of the metal, and the force to fracture (fracture force: N) was recorded; the deflection (mm) at fracture was also measured. Eight specimens were tested for each experimental condition, and the means and standard deviations were calculated. Statistical analysis of the results for the three-point bending test was performed using ANOVA/Duncan's Post Hoc multiple comparisons and t-test at a significance level of $\alpha = 0.05$.

Results

The results of the fracture force and deflection for the different thicknesses of each metal (d_M) are presented in Figs. 3 and 4, respectively.

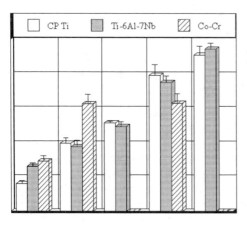 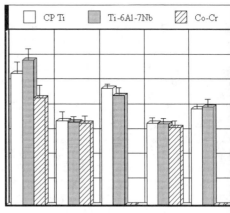

FIG. 3—*Fracture force of each metal with different thicknesses.*   FIG. 4—*Deflection at fracture.*

An increase in the thickness (d_M) of CP Ti and Ti-6Al-7Nb increased the fracture force. The specimens without α-case required greater fracture force compared to the specimens with α-case in all of the metals. There were no significant differences in the values (p<0.05) for Co-Cr alloy between the thicknesses of 0.5 and 0.7 mm. The deflection at fracture of the 0.3 mm-thick specimens was the highest for any of the metals. There were no statistical differences in deflection between the other two thicknesses (0.5 or 0.7 mm) for all of the metals, or among the metals for each of these thicknesses. The CP Ti and Ti-6Al-7Nb specimens without α-case deflected more than the specimens with α-case when the resin composite delaminated from the metal. The fractography conducted after the three-point bond test indicated that all of the specimens underwent adhesive fracture at the interface between the resin composite and the metal plate, and no cohesive failure was observed in the resin composite.

When the fracture force and deflection data for the specimens without α-case are not included in the statistical analysis, two-way ANOVA (two factors: metal and thickness) indicates statistical significance ($p < 0.01$) for each factor and the interactions of factors. The results of three-way ANOVA using the fracture force data for CP Ti and Ti-6Al-7Nb [$d_M = 0.5, 0.5$ (0.8-α), $0.7, 0.7$ (1.0-α)] are summarized in Table 2. Although there was no significance for the metal factor, the metal thickness and α-case factors indicated significances ($p < 0.001$).

TABLE 2 – *Summary of three-way ANOVA of fracture force for CP Ti and Ti-6Al-7Nb**

Source of variance	Sum of squares	df	Mean square	F	Sig.
Metal (a)	1.372	1	1.372	1.243	0.270
Thickness (b)	1641.161	1	1641.161	1486.792	<.001
α-case (c)	189.785	1	189.785	171.934	<.001
a * b	.654	1	0.654	0.593	0.445
a * c	3.549	1	3.549	3.215	0.078
b * c	3.827	1	3.827	3.467	0.068
a * b * c	3.739	1	3.739	3.388	0.071
Error	61.814	56	1.104		
Total	18640.235	64			

* Data used: $d_M = 0.5, 0.5$ (0.8-α), $0.7, 0.7$ (1.0-α).

Discussion

The fracture force of the CP Ti and Ti-6Al-7Nb determined by the three-point bond test was significantly higher after removal of the α-case compared to the specimens with α-case for two of the specimen thicknesses ($d_M = 0.5$ or 0.7 mm) (Fig. 3). Furthermore, the deflection at the fracture of the specimens without α-case was significantly greater than for the specimens with α-case (Fig. 4). These results indicate that the fracture force improved when the surface reaction layers were removed. In this study, all of the specimen surfaces were sandblasted. The reason for the decrease in the fracture force might be that the surface roughness itself is different on the surfaces with and without α-case after sandblasting. The surface of cast titanium is much harder compared to the bulk. Therefore, sandblasting the cast surface having the α-case is less effective at making a rough surface compared with sandblasting the polished surface without α-case. Sandblasting titanium requires a different technique (particle size and pressure) from the technique normally used for conventional dental casting alloys [23]. Suese et al. [15] reported that increasing the surface roughness by sandblasting with different-sized alumina particles increased the tensile bond strength of the adhesive resin to cast titanium. By increasing the roughness of the metal surface, sandblasting contributes to the formation of a mechanical bond. Using a shear bond test, Watanabe et al. [14] investigated the effect of surface contamination on the adhesive bonding of cast pure titanium and Ti-6Al-4V. They reported that the rough surface created by the reaction with the investment did not affect the mean shear bond strength of the resin cements, and that the surface roughness may have affected the tensile bond test results more than it affected the shear bond test results. In the three-point bending test used in the present study, the fracture force of the specimens with α-case was lower compared to the specimens without α-case. From this point of view, it is surmised that the fracture force determined by the three-point bond test was affected by the surface reaction layer and also that the three-point bond test may be more affected by surface roughness than the shear bond test.

It is interesting that no statistical differences in deflection were found between the 0.5 and 0.7 mm thicknesses for each metal or among all metals for each thickness. In the CP Ti and Ti-6Al-7Nb specimens with α-case, the force to fracture the resin composite on the 0.7-thick specimen was greater than on the 0.5-thick specimens, although they had similar deflections at fracture. This result indicates that the resin composite fractured at similar levels of deflection and the thicker specimens required higher force to deflect the metal plate to the same degree. A similar phenomenon happened for the CP Ti and Ti-6Al-7Nb specimens without α-case. On the other hand, the Co-Cr specimens required similar force for the thicknesses of 0.5 and 0.7 mm to deflect the metal at the point where the composite fractured, probably due to the stiffness of the Co-Cr alloy. The lowest fracture force occurred for the 0.3 mm-thick specimens. The 0.3 mm-thick CP Ti and Ti-6Al-7Nb specimens consisted of only the α-case. Regardless of the α-case, the strain at fracture for the 0.3 mm-thick specimens was the highest of any metal. The 0.3 mm-thick structure might be too thin to retain the veneer on the cast surfaces with the α-case even under a small load.

The results of this study suggested that the bond between a veneer and a thin metal framework ($d_M < 0.5$) may fail. Ideally, the surface reaction layer should be removed when preparing resin-veneered prostheses. However, eliminating the α-case from restorations has the potential to remove significant amounts of material and could affect the clinical adaptation of a prosthesis. The material loss resulting from removal of the α-case and from sandblasting is important to the clinical fit of restorations. Unnecessary grinding of the restorations should be avoided because it is likely to damage their margins and hamper the fit.

Acknowledgments

This study was supported by a Grant-in-Aid for scientific research (B) 11694293 and in part by a Grant-in-Aid for young scientists, (A) 11771229 from the Ministry of Education, Culture, Sports, Science & Technology, Tokyo, Japan. Support was also received from NIH/NIDCR grant DE11787.

References

[1] Bergman, B., Bessing, C., Ericson, G., Lundquist, P., Nilson, H., and Andersson, M. A., "2-Year Follow-Up Study of Titanium Crowns," *Acta Odontol Scand*, Vol. 48, 1990, pp.113–117.

[2] Gen, Y. and Tamaki, Y., "Grinding of Titanium with Experimental Sintered Diamond Wheels," *J Showa Dent*, Vol. 16, 1996, pp. 393–402.

[3] Ida, K., Tani, Y., Tsutsumi, S., Togaya, T., Nambu, T., and Suese, K., "Clinical Application of Pure Titanium Crowns," *Dent Mater J*, Vol. 4, 1985, pp. 191–195.

[4] Kononen, M., Rintanen, M., Waltimo, A., and Kempainen, P., "Titanium Framework

Removable Partial Denture Used for Patients Allergic to Other Metals," *J Prosthet Dent*, Vol. 73, 1995, pp. 4–7.

[5] Takahashi, J., Kimura, H., Lautenschlager, E. P., Chern Lin, J. H., Moser, J. B., and Greener, E. H., "Casting Pure Titanium into Commercial Phosphate-Bonded SiO_2 Investment Molds," *J Dent Res*, Vol. 69, 1990, pp. 1800–1805.

[6] Miyakawa, O., Watanabe, K., Okawa, S., Nakano, S., Kobayashi, M., and Shiokawa, N., "Layered Structure of Cast Titanium Surface," *Dent Mater J*, Vol. 8, 1986, pp. 175–185.

[7] Watanabe, I., Watkins, H., Nakajima, H., Atsuta, M., and Okabe, T., "Effect of Pressure Difference on the Quality of Titanium Casting," *J Dent Res*, Vol. 76, 1997, pp.773–779.

[8] Ida, K., Togaya, T., Tsutsumi, S., and Takeuchi, M., "Effect of Magnesia Investments in the Dental Casting of Pure Titanium and Titanium Alloys," *Dent Mater J*, Vol. 1, 1982, pp. 8–21.

[9] Watari, F., "High Temperature Reactivity Between Titanium and Refractory Oxides in Dental Casting of Procedure," *J Jpn Dent Mater*, Vol. 8, 1989, pp. 83–96.

[10] Taira, M., Moser, J. B., Greener, E. H., "Studies of Ti Alloys for Dental Casting," *Dent Mater*, Vol. 5, 1989, pp. 45–50.

[11] Vallittu, P. K. and Kokkonen, M., "Deflection Fatigue of Cobalt-Chromium, Titanium, and Gold Alloy Cast Denture Clasp," *J Prosthet Dent*, Vol. 74, 1995, pp. 412–419.

[12] Dummer, P. and Gidden, J., "The Maryland Bridge: A Useful Modification," *Journal of Dentistry*, Vol. 14, 1986, pp. 42–43.

[13] Watanabe, I., Matsumura, H., and Atsuta, M., "Effect of Two Metal Primers on Adhesive Bonding with Type IV Gold Alloys," *J Prosthet Dent*, Vol. 73, 1995, pp. 299–303.

[14] Watanabe, I., Watanabe, E., Yoshida, K., and Okabe, T., "Effect of Surface Contamination on Adhesive Bonding of Cast Pure Ti and Ti-6Al-4V," *J Prosthet Dent*, Vol. 81, 1999, pp. 270–276.

[15] Suese, K., Iwai, K., Kakuta, J., Ohtsuka, K., Sakaida, F., and Kawazie, T., "Prosthodontic Evaluation on Casting Restoration of Pure Titanium," *J Jpn Prosth Dent*, Vol. 28, 1984, pp. 860–867.

[16] Tanaka, T., Atsuta, M., Nakabayashi, N., and Masuhara, E., "Surface Treatment of Gold Alloys for Adhesion," *J Prosthet Dent*, Vol. 60, 1988, pp. 271–279.

[17] Schwickerath, H., "Zur verbundfestigkeit von metallkeramik," *Dtsch Zahnarztl Z*, Vol. 35, 1980, pp. 910–912.

[18] Deutsches Institute Fur Normung, DIN 13927, Metall-Keramik-Systeme, Beuth Verlag, Berlin, 1990.

[19] Lenz, J., Schwarz, S., Schwickerath, H., Spermer, F., and Schafer, A., "Bond Strength of Metal-Ceramics in Three-Point Flexure Bond Test," *J Applied Biomaterials*, Vol. 6, 1995, pp. 55–64.

[20] Pang, I., Gilbert, J., Chai, J., and Lautenschlager, E., "Bonding Characteristics of Low-Fusing Porcelain Bonded to Pure Titanium and Palladium-Copper Alloy," *J Prosthet Dent*, Vol. 73, 1995, pp. 17–25.

[21] Probster, L., Maiwald, U., and Weber, H., "Three-Point Bending Strength of Ceramics Fused to Cast Titanium," *Eur J Oral Sci*, Vol. 104, 1996, pp. 313–319.

[22] Schwarz, S., Lenz, J., and Schwickerath, H., "Zur Festigkeit des metallkeramischen verbundes bei der biegeprufung," *Dtsch Zahnarztl Z*, Vol. 43, 1988, pp. 1152–1158.

[23] Kurtz, K., Kabcenell, J., Watanabe, I., and Okabe, T., "Shear Bond Strength of Polymer-Glass Composite to Cast Titanium," *J Dent Res*, Vol. 77, 272, #1331, 1998.

Journal of ASTM International, October 2005, Vol. 2, No. 9
Paper ID JAI12783
Available online at www.astm.org

Yoshimitsu Okazaki[1] *and Emiko Gotoh*[2]

Corrosion Resistance, Mechanical Properties, Fatigue Properties, and Tissue Response of Ti-15Zr-4Nb-4Ta Alloy

ABSTRACT: Ti, Zr, Nb, and Ta are biocompatible elements. A Ti-15Zr-4Ta-4Nb alloy for medical implants is being developed. In terms of in vitro cytocompatibility, the new bone tissue response to the Ti-15Zr-4Ta-4Nb alloy determined through rat tibia implantation was equal to or better than that to the Ti-6Al-4V alloy. An acceptable level of biological response can be expected when this alloy is used appropriately because it consists of biocompatible elements and has excellent biocompatibility, corrosion resistance, microstructure, and mechanical and fatigue properties. Therefore, the Ti-15Zr-4Ta-4Nb alloy with a low metal release rate is considered advantageous for long-term surgical implants.

KEYWORDS: corrosion resistance, metal release, mechanical properties, fatigue strength, rat tibia implantation, bone tissue response

Introduction

Ti, Zr, Nb, and Ta exhibit excellent biocompatibility, and belong to the loose connective vascularized (vital) group with regards to tissue reaction (Fig. 1) [1]. It has been reported that of the 70 metals in the periodic table, only Zr and Ti support osteoblast growth and osteointegration [2]. Ti, Zr, Nb and Ta also have a considerably superior corrosion resistance. The effects of various metals on cell viability have been reported using metallic particles [3–4]. Since the quantities of metal released into the medium were small (<0.3 μg/mL) for Ti, Zr, Ta, and Nb particle extractions, the relative growth ratios of murine fibroblast L929 and murine osteoblastic MC3T3–E1 cells were equal to one (noncytotoxic).

The effects of alloying elements on the Ti alloy structure and the Ti alloys specified by the ASTM International and the Japanese Industrial Standard (JIS) are summarized in Fig. 2. Many (α–β)–type Ti alloys consisting of a mixed alpha (hcp: hexagonal close–packed crystal)–beta (bcc: body centered cubic crystal) structure have been standardized by ASTM International. The near β–type Ti-13Nb-13Zr alloy, and β–type Ti-15Mo and Ti-12Mo-6Zr-2Fe alloys, which have Young's moduli slightly lower than that of the (α–β)- type alloy, are specified in the ASTM standard. Ti-15Zr-4Nb-4Ta has been standardized in the JIS.

Our research group previously reported on the effects of Zr, Nb, Ta, and Pd on the mechanical properties of Ti alloys and the biocompatibilities of these alloys to cultured cells, and on the corrosion resistance of Ti alloys, as determined using anodic polarization tests in synthetic body fluids [5–13]. If the current density measured by anodic polarization testing is low, the corrosion resistance of the metallic biomaterial is excellent. The anodic polarization property of Ti alloy is improved when Zr, Nb, Ta, and Pd are added because the resultant ZrO_2, Nb_2O_5,

Manuscript received 13 August 2004; accepted for publication 26 January 2005; published October 2005.
Presented at ASTM Symposium on Titanium, Niobium, Zirconium, and Tantalum for Medical and Surgical Applications on 9-10 November 2004 in Washington, DC.
[1] Institute for Human Science and Biomedical Engineering, National Institute of Advanced Industrial Science and Technology, 2–1 Namiki 1–chome, Tsukuba, Ibaraki 305-8564, Japan.
[2] National Institute of Technology and Evaluation, 2 Namiki 1–chome, Tsukuba, Ibaraki 305-0044, Japan.

Ta$_2$O$_5$ and PdO strengthen the TiO$_2$ passive film that formed on the Ti alloy. The anodic polarization property under friction is also excellent [14]. The Ti alloy disk is worn with an apatite ceramic pin in Eagle's medium, and the wear powder is sterilized in ethanol and added to the culture medium [15]. The growth ratios of L929 and MC3T3–E1 cells with the Ti-6Al-4V alloy wear powder relative to that of the control cells are lower than those of the L929 and MC3T3–E1 cells with the Ti-15Zr-4Nb-4Ta alloy wear powder.

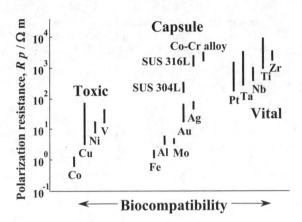

FIG. 1—*Relationship between polarization resistance and biocompatibility of pure metal and alloy reported by Steinmann [1]. Zr, Nb, and Ta are biocompatible elements and improve corrosion resistance.*

Alpha structure (Hcp)	Mixed alpha-beta structure	Beta structure (Bcc)
◄——— Alpha-stabilizing elements [Al, O, Zr etc.]		Beta-stabilizing elements ———► [Mo, Fe, Nb, Ta etc.]
	◄——— Higher fatigue strength	
	Improved fabricability ———►	
Unalloyed Titanium (ASTM F 67)	Ti-6Al-4V (ASTM F 136) Ti-3Al-2.5V (ASTM F 2146) Ti-6Al-7Nb (ASTM F 1295) Ti-13Zr-13Nb (Near β) (ASTM F 1713) Ti-6Al-2Nb-1Ta (JIS T 7401-3) Ti-15Zr-4Nb-4Ta (JIS T 7401-4)	Ti-15Mo (ASTM F 2066) Ti-12Mo-6Zr-3Fe (ASTM F 1813) Ti-15Mo-5Zr-3Al (JIS T 7401-6)

FIG. 2—*Effects of alloying elements on Ti alloy structure, and Ti alloy types specified in ASTM and JIS.*

The concentration of V released from the wear powder into the medium increases with increasing amount of wear powder. For the Ti-15Zr-4Nb-4Ta wear powder, the maximum Ti concentration released from the wear powder roughly agrees with the results obtained using high-purity Ti particles [15]. In contrast, for Zr, Nb, and Ta, the maximum metal concentrations released from the wear powders are much lower than those obtained using high–purity metal particles. The mechanical strength at room temperature of the new Ti alloy annealed at 700°C for 2 h is increased by adding Zr and small quantities of oxygen and nitrogen [5,6,10]. The relative growth ratios of L929 and MC3T3–E1 cells are estimated using the following formula: (average number of cells per dish after 4 d incubation)/(average number of cells in the control). The relative growth ratios of the L929 (1.09±0.04) and MC3T3–E1 (1.08±0.02) cells with the Ti-15Zr-4Nb-4Ta alloy are slightly higher than those of L929 and MC3T3–E1 cells with the Ti-6Al-4V alloy [11,12]. In this study, the corrosion resistance, mechanical properties, and fatigue properties of and bone tissue response to Ti-15Zr-4Nb-4Ta alloy were investigated.

Materials and Methods

Alloy Specimens

The Ti-6Al-4V extra low interstitial (ELI) and Ti-15Zr-4Nb-4Ta alloys were subjected to vacuum-arc melting. After β (after soaking: 1150°C–3 h for Ti-6Al-4V and 1050 °C–4 h for Ti-15Zr-4Nb-4Ta) and α–β forging (starting temperature: 930°C for Ti-6Al-4V and 750°C for Ti-15Zr-4Nb-4Ta), the Ti alloys were annealed for 2 h at 700°C. In addition, to determine the optimum solution treatment and aging conditions for the Ti-15Zr-4Nb-4Ta alloy, the specimens were kept between 755 and 800°C for 1 h and then cooled in water. After the solution treatment, the specimens were aged between 350 and 450°C for 5–15 h and then cooled in air. Table 1 shows the chemical compositions of the Ti alloys.

TABLE 1—*Chemical compositions (mass%) of specimens.*

Titanium Alloy	Zr	Nb	Ta	Pd	Al	V	Fe	O	N	H	C	Ti
Ti-15Zr-4Nb-4Ta	14.83	3.97	4.01	0.16	-	-	0.04	0.22	0.05	0.0057	0.01	Bal.
Ti-6Al-4V	-	-	-	-	6.40	4.40	0.10	0.07	0.02	0.0027	0.025	Bal.

Tensile Tests and Microscopic Observation

Tensile tests were conducted using test specimens 6 mm in diameter and 22 mm in gauge length at room temperature and a crosshead speed of 0.5 mm/min. The test specimens were etched and then their microstructures were observed using optical microscopy and transmission electron microscopy (TEM).

Immersion Test

The immersion test was performed in accordance with the JIS T 0304 standard for metallic biomaterials. Plate specimens, each 20 mm × 40 mm × 1 mm, were cut from each alloy specimen. Immersion tests were conducted at 37°C using eight solutions, namely, (1) α–medium (containing NaCl, 6.8 g; KCl, 0.4 g; Na_2HPO_4, 1.15 g; $NaH_2PO_4H_2O$, 0.2 g/L; and trace amounts of amino acids and vitamins, pH = 7.4); α–medium containing 10 vol% fetal bovine serum and 7.5% $NaHCO_3$ solution (1 vol%); (2) phosphate–buffered saline [PBS(–), containing NaCl, 8 g; KCl, 0.2 g; NaH_2PO_4, 0.14 g, and KH_2PO_4, 0.2 g/L, pH = 7.2); (3) membrane-filtered calf serum (pH = 6.9), (4) 0.9 mass% NaCl (pH = 6.6); (5) artificial saliva [containing NaCl, 0.844 g; KCl,

1.2 g; $CaCl_2$, 0.146 g; $MgCl_2$, 0.052 g; K_2PO_4, 0.342 g/L; and small amounts of thickener (carboxy methyl cellulose), and conservative medium (sodium benzoate and sorbic acid), pH = 6.2, viscosity: 6 mPas]; (6) 1 mass% lactic acids (pH = 2.6); (7) 1.2 mass% L–cysteine (pH = 2.1)–simulated amino acid; and (8) 0.05 vol% concentrated HCl (0.01 mass% HCl, pH = 2.0). The 1.2 mass% L-cysteine solution was prepared from L-cysteine hydrochloride monohydrate. The 0.05 vol% concentrated HCl solution was prepared from ultra–high-purity hydrochloric acid concentrate. The plate specimens were surface-finished with waterproof emery paper up to 1000 grit under running water, and then ultrasonically cleaned. These specimens were placed in polypropylene bottles and separately sterilized in an autoclave at 121°C for 30 min. The PBS(–) was sterilized in an autoclave. The other solutions, except for the artificial saliva and 0.01 % HCl, were sterilized by passing them through a 0.2–μm-pore-size membrane filter. The artificial saliva was used as received because it contained small amounts of thickener and conservative medium. Then 50 mL of each solution was poured into polypropylene bottles, each containing a plate specimen. All the bottles were placed inside an incubator at 37°C for 7 days.

The concentrations of various metals released into solution were determined in ppb (ng/mL) by inductively coupled plasma-mass spectrometry (ICP-MS). A solution without a metal specimen was incubated under similar conditions and used for the blank test. The quantity of metal released ($\mu g/cm^2$) was estimated using the following formula: (amount of solution: 50 mL) × [(metal concentration in each test solution) - (mean metal concentration in blank test with three bottles)]/(surface area of specimen). The mean quantity of metal released and standard deviation were calculated for five bottles.

Fatigue Tests

The fatigue test was performed in accordance with the ASTM F 1801 standard. Fatigue test specimens with the shapes and dimensions shown in Fig. 3 were cut from the sample alloys. Each specimen was fitted into a polyethylene-testing cell containing Eagle's medium, and then set on a fatigue-testing machine [17]. Fatigue tests were conducted at an internal cell temperature of 37°C, maintained by circulating heated water around the cell. A small amount of gas containing 90 % N_2, 5 % CO_2 and 5 % O_2 was bubbled into the solution. Test conditions were set at a stress ratio [R = (minimum stress)/(maximum stress)] of 0.1, a frequency of 10 or 2 Hz, and a maximum of 10^9 cycles. Additionally, fatigue tests were conducted using a hip joint load profile, estimated by an analysis of the movement of the human hip joint and the forces acting upon it (hip joint load profile) [18]. Moreover, the fatigue test under the torsion-to-torsion mode was carried out with a sine wave of R = 0.1 and a frequency of 10 Hz.

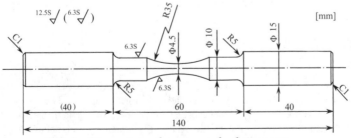

FIG. 3—*Dimensions of specimens for fatigue test.*

Rat Tibia Implantation

Implant specimens, 1.2 mm in diameter and 2 mm in length, were cut from the alloy specimens. The surface roughnesses (Ras) of the Ti-6Al-4V and Ti-15Zr-4Nb-4Ta alloys were 1.36 ± 0.18 and 0.77 ± 0.07, respectively.

Six-week-old male Wistar rats weighing 184 ± 30 g were selected. For the implant cavity, two holes 1.2 mm in diameter were formed 8 and 13 mm below the knee joint. Two of the same alloy implants in a limb were anteroposteriorly bicortically inserted and not positioned inside the medullary cavity parallel to the longitudinal axis of the tibia. The Ti-6Al-4V and Ti-15Zr-4Nb-4Ta alloys were implanted separately into the left or right tibia for 6, 12, 24 and 48 weeks. Five Wistar rats without implants were raised under similar conditions and were used as a control group.

Bone tissue was fixed in a 10 % formalin solution for 3 days. Tissues embedded in resin were cut with a diamond blade to a thickness of approximately 100 μm parallel to the length of the tibia to divideit into two the vertical axis of the Ti alloy implant. The sliced resin block was then polished to a thickness below 40 μm with waterproof emery paper, followed by staining with toluidine blue (T. B.) and Villanueva stains. After fixation, the other bone tissue was decalcified with a 5 % formalin solution containing 5 % formic acid (pH = 2.3, 10 % formalin: 10 % formic acid = 1:1) for 3 days. The bone tissue was dehydrated and defatted with alcohol (70, 80, 90, 100 %) and 100 % acetone successively. The specimens embedded in paraffin wax were sliced to a thickness of approximately 4 μm. These sliced sections were stained with hematoxylin eosin (H. E.) and azan malloy (A. M.) stains. The new bone tissues that formed around the Ti implant were compared using the following four parameters, as estimated using image analysis: bone formation rate (%) = (total circumference of new bone formed around Ti implant)/ (circumference of implant); bone contact rate (%) = (total circumference in contact with implant as determined from ×100 optical micrograph)/(circumference of implant); thickness of new bone = (total area of new bone)/(total circumference of new bone formed around Ti implant); and bone maturity (%) using A. M. or Villanueva stain = (total area of mature bone)/(total area of new bone).

After decalcification, the tibia tissues were crosscut at two sites nearly 2.5 mm from the center of the implants. The implants were carefully removed, and then lyophilized with a freeze dryer. The lyophilized tibia tissues were dissolved by microwave acid digestion in a sealed Teflon decomposition vessel [16]. The metal concentration in the solution containing the dissolved tibia tissues was determined by ICP-MS. The concentrations (μg/g) of various metals in the tibia tissues were calculated by dividing the quantity of each metal (μg) in the dissolved tibia tissue solution by the lyophilized tibia tissue weight (g). The means and standard deviations were calculated from results obtained from five rats.

Results and Discussion

In Vitro and In Vivo Metal Releases

The quantity of Ti released from the Ti-15Zr-4Nb-4Ta alloy was much smaller than that released from the Ti-6Al-4V alloy in all the solutions (Fig. 4). The quantities of alloying elements (Zr, Nb, and Ta) released were also very low. In particular, the quantity of Ti or alloying elements released from the Ti alloy markedly increased with decreasing pH (pH<4), and it markedly attenuated at pHs of approximately four and higher. The ratio of the average quantity of Ti released from the Ti-15Zr-4Nb-4Ta alloy to that released from the Ti-6Al-4V alloy was calculated using the following equation: ratio of Ti released = (average quantity of Ti released from Ti-15Zr-4Nb-4Ta)/(average quantity of Ti released from Ti-6Al-4V). The ratio of Ti released from the Ti-15Zr-4Nb-4Ta alloy was smaller than that from the Ti-6Al-4V alloy in all the solutions (Fig. 5a). In particular, it was approximately 30 % or smaller in 1 % lactic acid, 1.2 % cysteine and 0.01 % HCl.

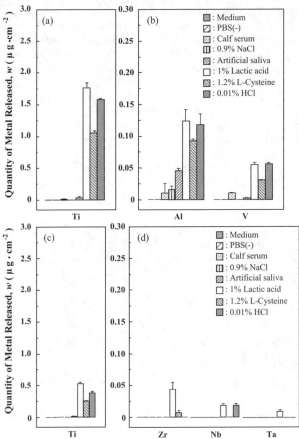

FIG. 4—*Quantity of metal released into various solutions at 37°C after one week: (a) Ti released from Ti-6Al-4V ELI alloy, (b) Al and V released from Ti-6Al-4V ELI alloy, (c) Ti released from Ti-15Zr-4Nb-4Ta alloy, and (d) Zr, Nb, and Ta released from Ti-15Zr-4Nb-4Ta alloy.*

The ratio of alloying element released = (sum of average quantities of each alloying element released from Ti-15Zr-4Nb-4Ta alloy) / (sum of average quantities of Al and V released from Ti-6Al-4V alloy) was also calculated. The quantity of (Zr+Nb+Ta) released was also considerably lower than that of (Al+V) released.

The Ti concentration in the tibia tissues with the Ti-15Zr-4Nb-4Ta implant was considerably lower than that in the tibia tissues with the Ti-6Al-4V implant (Fig. 6). The Zr, Nb and Ta concentrations in the tibia tissues with the Ti-15Zr-4Nb-4Ta implant did not increase markedly compared with those of the control group. The ratio of Ti concentration or the total concentration of each alloying element in the tibia tissues with the Ti-15Zr-4Nb-4Ta implant to that in the tibia tissues with the Ti-6Al-4V implant was estimated. The Ti concentration in the tibia tissues with the Ti-15Zr-4Nb-4Ta implant was approximately 40 % lower than that in the tibia tissues with the Ti-6Al-4V implant (Fig. 5b). The decrease in Ti ratio with increasing implantation period was caused by both the increase in the quantity of Ti released from the Ti-6Al-4V implant and the decrease in the quantity of Ti released from the Ti-15Zr-4Nb-4Ta implant as shown in Fig. 6a. The (Zr+Nb+Ta) concentration was 20 % less than the (Al+V) concentration.

Since metal release from implants is an important subject, numerous studies, including long-term clinical studies, have been conducted on metal release into the body from orthopedic implants [19–24]. The toxic effects of metals released from prosthetic implants have been reviewed [19]. Metals from orthopedic implants are released into surrounding tissue by various mechanisms, including corrosion, wear, and mechanically accelerated electrochemical processes such as stress corrosion, corrosion fatigue and fretting corrosion. This metal release has been associated with clinical implant failure, osteolysis, cutaneous allergic reactions, and remote site accumulation [23,24]. The increase in the incidence of allergy and the necessity for prolonged use require implants with minimal metal release rate. Therefore, the Ti-15Zr-4Nb-4Ta alloy with a much lower metal release rate is advantageous for use in long-term implants.

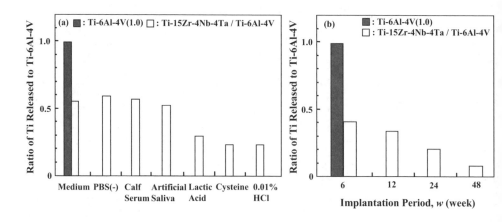

FIG. 5—*Ratios of Ti ions released from Ti-15Zr-4Nb-4Ta alloy to that released from Ti-6Al-4V ELI alloy (Control: 1.0) in vitro and in vivo. (a) Immersion test for 7 days; (b) Ti concentration in rat bone tibia tissue with implants.*

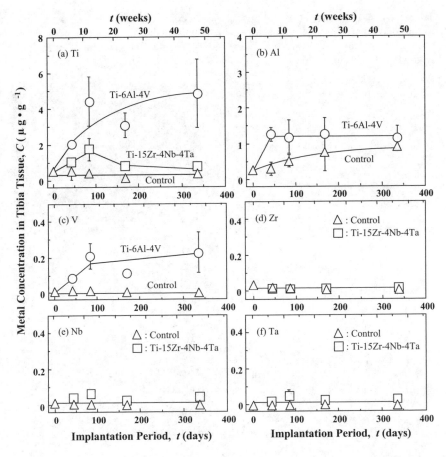

FIG. 6—*Change in metal concentration in lyophilized tibia tissue with Ti-6Al-4V ELI or Ti-15Zr-4Nb-4Ta implants as function of implantation period. (a) Ti released from Ti-6Al-4V ELI or Ti-15Zr-4Nb-4Ta; (b) Al released from Ti-6Al-4V ELI; (c) V released from Ti-6Al-4V ELI; (d) Zr released from Ti-15Zr-4Nb-4Ta; (e) Nb released from Ti-15Zr-4Nb-4Ta; and (f) Ta released from Ti-15Zr-4Nb-4Ta.*

Microstructure and Mechanical Properties

From the optical and TEM observations, the Ti-15Zr-4Nb-4Ta alloy annealed at 700°C for 2 h was shown to consist mostly of laths α'-martensite, whereas the solution-treated alloy consisted of the primary α-phase as well as α'-martensite (Fig. 7). The fine α-phase precipitate was due to the aging after solution treatment. A good balance between strength and ductility was obtained when the fine α-phase was uniformLy distributed as a result of solution treatment combined with aging. We assumed that the optimum heat treatment conditions for the Ti-15Zr-4Nb-4Ta alloy were solution treatment (S.T.) at 775°C for 1 h followed by aging at 400°C for

8 h. The mechanical properties at room temperature are compared in Table 2. The relationships between the total elongation and ultimate tensile strength of the Ti alloys (mainly annealed Ti alloys) the specified in ASTM and the Ti-15Zr-4Nb-4Ta alloy are compared in Fig. 8. The ductility and tensile strength of the Ti-15Zr-4Nb-4Ta alloy were high.

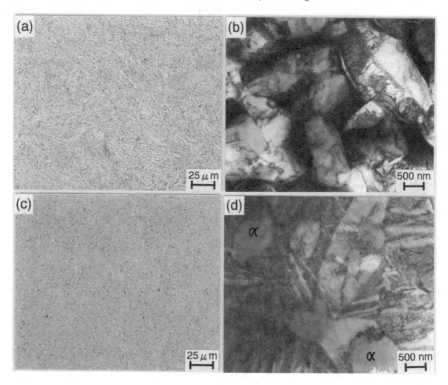

FIG. 7— *Optical and TEM images of Ti-15Zr-4Nb-4Ta alloy annealed at 700°C for 2 h (a and b), and aged at 400°C for 8 h after solution treatment at 775°C for 1 h (c and d).*

TABLE 2—*Comparison of mechanical properties of alloys at room temperature.*

		$\sigma_{0.2\%PS}$ (MPa)	σ_{UTS} (MPa)	T.E.(%)	R.A.(%)
Ti-15Zr-4Nb-4Ta	Annealed	908	945	21	61
	S.T. + Aged	1045	1156	19	65
Ti-6Al-4V	Annealed	849	936	15	42

σ 0.2%PS : 0.2% Proof strength, σ UTS : Ultimate tensile strength, T. E.: Total elongation,
R. A.: Reduction of area, S.T. + Aged : Aged at 400 °C for 8 h after solution treatment at 775 °C for 1 h

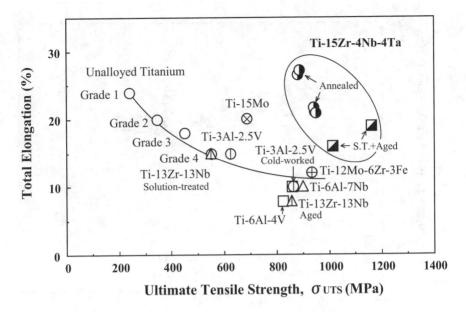

FIG. 8—*Relationships between minimum ultimate tensile strengths and total elongation values of Ti alloys specified in ASTM and Ti-15Zr-4Nb-4Ta alloy.*

Fatigue Strength

The number of cycles required to induce the failure of the Ti-15Zr-4Nb-4Ta alloy annealed at 700°C increased as the maximum stress decreased. It was found that rupture stress at 1×10^8 cycles occurred at approximately 600 MPa (Fig. 9). The S–N curves for loading with a sine wave or a hip joint load profile were also similar. The fatigue strength of the (α-β)-type Ti-15Zr-4Nb-4Ta alloy having the α'-martensite markedly increased compared to that of the β-type Ti alloy with a single bcc structure [10,25]. The fatigue strength after solution treatment plus aging was higher than that after annealing, and was the same under sine wave and hip joint profile loadings. The ratio of the fatigue strength to ultimate tensile strength of the Ti-15Zr-4Nb-4Ta alloy was approximately 65 %, regardless of whether it had been aged after solution treatment or annealed. Moreover, the fatigue strength under the torsion-to-torsion mode was also high (Fig. 10).

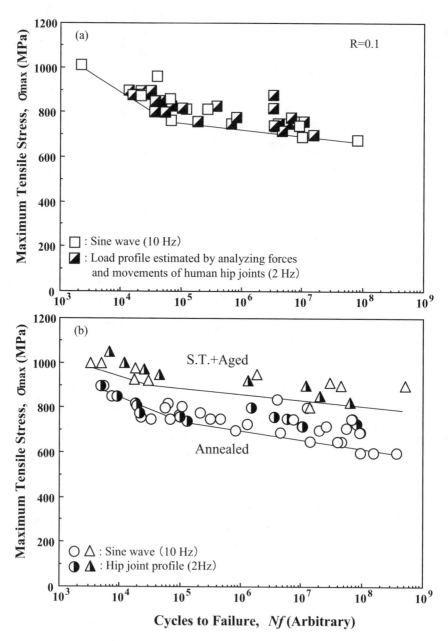

FIG. 9— *Comparison of S–N curves obtained by tensile fatigue tests with sine wave or hip joint load profiles in Eagle's medium at 37°C for Ti-6Al-4V ELI(a) and Ti-15Zr-4Nb-4Ta alloys (b).*

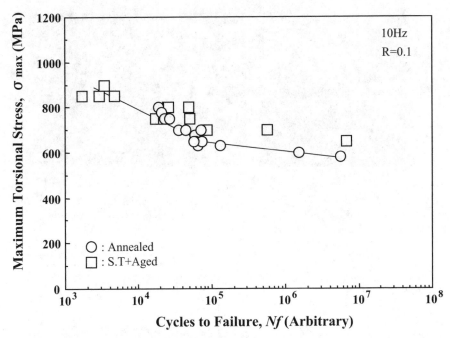

FIG. 10— *Comparison of S-N curves obtained by torsional fatigue tests with sine wave in Eagle's medium at 37°C for Ti-15Zr-4Nb-4Ta alloy.*

Bone Tissue Response

The Ti-15Zr-4Nb-4Ta alloy implant was completely surrounded by new bone. No foreign-body giant cells or inflammatory responses were observed (Fig. 11). New bone formed similarly around both the Ti-6Al-4V and Ti-15Zr-4Nb-4Ta alloys implanted in the bone marrow (Fig. 12). Statistical analyses showed no significant differences ($p < 0.05$) in the thicknesses of new bone tissues 6, 12, 24, and 48 weeks after the implantation of both alloys.

Practicability

The facilities required for the melting, forging, and working processes of the Ti-15Zr-4Ta-4Nb alloy are the same as those required for those of the Ti-6Al-4V alloy. Bone plates, hip screws, intramedullary fixation, and artificial hip joint implants are fabricated by the same conventional manufacturing processes used for the Ti-6Al-4V alloy (Fig. 13).

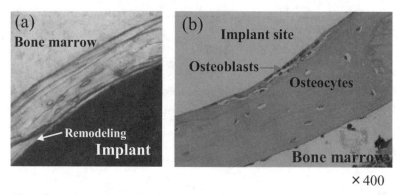

FIG. 11— *Optical micrograph of nondecalcified bone stained with toluidine blue and surrounding the Ti-15Zr-4Nb-4Ta implant 48 weeks after implantation (a); and that decalcified bone stained with H. E. 12 weeks after implantation (b).*

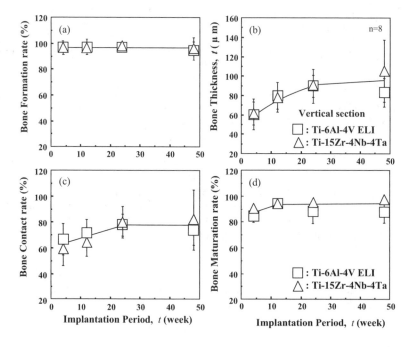

FIG. 12— *Change of newly formed bone around implants as function of implantation period: (a) bone formation rate; (b) bone thickness; (c) bone contact rate; and (d) bone maturation rate.*

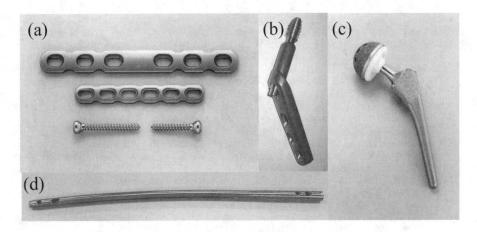

FIG. 13—*Bone plate (a), compression hip screw (b), cementless artificial hip joint (c), and intramedullary fixation (d) made of Ti-15Zr-4Nb-4Ta alloy.*

Conclusions

In vitro and in vivo metal releases from the Ti-15Zr-4Nb-4Ta alloy is much lower than that from the Ti-6Al-4V extra low interstitial alloy. The ductility and tensile strength of the Ti-15Zr-4Nb-4Ta alloy were higher than those of the Ti-6Al-4V alloy. The fatigue strength of the Ti-15Zr-4Ta-4Nb alloy annealed at 700°C for 2 h was approximately 600 MPa in Eagle's medium at 1×10^8 cycles, and this increased to 800 MPa after solution treatment plus aging. The ratio of fatigue strength to tensile strength was high at approximately 65 %. The fatigue strength under the torsion-to-torsion mode was also high. The Ti-15Zr-4Nb-4Ta alloy implant was completely surrounded by newly formed bone. The new bone tissue response to the Ti-15Zr-4Ta-4Nb alloy determined through rat tibia implantation was equal to or better than that to the Ti-6Al-4V alloy. The Ti-15Zr-4Nb-4Ta alloy with its excellent mechanical properties, corrosion resistance, fatigue properties, and ability to bone tissue response can be used for medical purposes.

References

[1] Steinemann, S. G., "Corrosion of Surgical Implants—In Vivo and In Vitro Tests," *Evaluation of Biomaterials, Advances in Biomaterials,* Chichester, Wiley, G. D. Winter, J. L Leray, K. de Goot, Eds., 1980, pp. 1–34.

[2] Steinemann, S. G., "Compatibility of Titanium in Soft and Hard Tissue—The Ultimate is Osseointegration," *Materials for Medical Engineering, Euromat 99,* Weinheim, Wiley–VCH, H. Stallforth, P. Revell, Eds., Vol. 2, 1999, pp. 199–203.

[3] Okazaki, Y., Rao, S., Asao, S., Tateishi, T., Katsuda, S., and Furuki, Y., "Effects of Ti, Al and V   Concentrations on Cell Viability," *Mater Trans JIM,* Vol. 39, 1998, pp. 1053–1062.

[4] Okazaki, Y., Rao, S., Asao, S., Tateishi, T., "Effects of Metallic Concentrations Other than Ti, Al and V on Cell Viability," *Mater Trans JIM,* Vol. 39, 1998, pp. 1070–1079.

[5] Ito, A., Okazaki, Y., Tateishi T., and Ito, Y., "In Vitro Biocompatibility, Mechanical Properties, and Corrosion Resistance of Ti-Zr-Nb-Ta-Pd and Ti-Sn-Nb-Ta-Pd Alloys," *J Biomed Mater Res,* Vol. 29, 1995, pp. 893–900.

[6] Okazaki, Y., Ito, Y., Ito A., and Tateishi, T., "New Titanium Alloys to be Considered for Medical Implant," *Medical Applications of Titanium and Its Alloys: The Material and Biological Issues, ASTM STP 1272*, S. A. Brown and J. E. Lemons, Eds. ASTM International, West Conshohocken, 1996, pp. 45–59.

[7] Okazaki, Y., Tateishi, T., and Ito, Y., "Corrosion Resistance of Implant Alloys in Pseudo Physiological Solution and Role of Alloying Elements in Passive Films," *Mater Trans JIM*, Vol. 38, 1997, pp.78–84.

[8] Okazaki, Y., Kyo, K., Ito, Y., and Tateishi, T., "Effects of Mo and Pd on Corrosion Resistance of V–Fee Titanium Alloys for Medical Implants," *Mater Trans JIM*, Vol. 38, 1997, pp. 344–352.

[9] Okazaki, Y., Tateishi, T., and Ito, Y., "Corrosion Resistance of Implant Alloys in Pseudo Physiological Solution and Role of Alloying Elements in Passive Films," *Mater Trans JIM*, Vol. 38, 1997, pp. 78–84.

[10] Okazaki, Y., Rao, S., Ito, Y., Tateishi, T., "Corrosion Resistance, Mechanical Properties, Corrosion Fatigue Strength and Cytocompatibility of New Ti alloys Without Al and V," *Biomaterials*, Vol. 19, 1998, pp. 1197–1215.

[11] Okazaki, Y. and Ito, Y., "New Ti Alloys Without Al and V for Medical Implant," *Advanced Eng Mater*, Vol. 2, 2002, pp. 278–281.

[12] Okazaki, Y., "A New Ti-15Zr-4Nb-4Ta Alloy for Medical Applications," *Current Opinion in Solid State and Materials Science*, Vol. 5, 2001, pp. 45–53.

[13] Okazaki, Y. and Gotoh, E., "Implant Applications of Highly Corrosion-resistant Ti-15Zr-4Nb-4Ta Alloy," *Mater Trans JIM*, Vol. 43, 2002, pp. 2943–2948.

[14] Okazaki, Y., "Effect of Friction on Anodic Polarization Properties of Metallic Biomaterials," *Biomaterials*, Vol. 23, 2002, pp. 2071–2077.

[15] Okazaki, Y. and Nishimura, E., "Effect of Metal Released from Ti Alloy Wear Powder on Cell Viability," *Mater Trans JIM*, Vol. 41, 2000, pp. 1247–1255.

[16] Okazaki, Y., Nishimura E., Nakada H., Kobayashi K.,"Surface Analysis of Ti-15Zr-4Nb-4Ta Alloy after Implantation in Rat Tibia," *Biomaterials*, Vol. 22, 2001, pp. 599–607.

[17] Okazaki, Y., Gotoh, E., "Corrosion Fatigue Properties of Metallic Biomaterials in Eagle's Medium," *Mater Trans JIM*, Vol. 43, 2002, pp. 2943–1955.

[18] Mejia, L. C. and Brierley, T. J., "A Hip Wear Simulator for the Evaluation of Biomaterials in Hip Arthroplasty Components, "*Bio-Medical Mater and Eng*, Vol. 4, 1994, pp. 259–271.

[19] Wapner, K L., "Implications of Metallic Corrosion in Total Knee Arthroplasty," *Clinical Orthopedics and Related Research*, Vol.271, 1991, pp. 12–20.

[20] Pazzaglia, U. E., Minoia, C., Gualtieri, G., Gualtieri, I., Riccardi, C., and Ceciliani, L., "Metal Ions in Body Fluids after Arthroplasty," *Acta Orthop Scand* , Vol. 57, 1986, pp. 415–418.

[21] Jacobs, J. J, Skipor, A. K., Black, J., Urban, R. M., and Galante, J. O., "Release and Excretion of Metal in Patients who Have a Total Hip-Replacement Component Made of Titanium–base Alloy," *J Bone Joint Surg* , Vol. 73–A, 1991, pp. 1475–1486.

[22] Jacobs, J. J, Skipor, A. K., Patterson, L. M., Hallab, N. J., Paprosky, W. G., Black, J., and Galante J. O., "Metal Release in Patients who Have had a Primary Total Hip Arthroplasty," *J Bone Joint Surg*, Vol. 80–A, 1998, pp.1447–1458.

[23] Jacobs, J. J, Silverton, C., Hallab, N. J., Skipor, A. K., Patterson, L., Black J., and Galante J. O., "Metal Release and Excretion from Cementless Ttitanium Alloy Total Knee Replacements," *Clinical Orthopedics and Related Research*, 1999, 358, pp.173–180.

[24] Agins, H. J, Alcock N. W., Bansal, M., Salvati, E. A., Wilson, Jr. P. D., Pellicci, P. M., and Bullough P. G., "Metallic Wear in Failed Titanium–Alloy Total Hip Replacements," *J Bone Joint Surg*, Vol. 70–A, 1988, pp. 347–356.

[25] Bhambri, S. K., Shetty, R. H., and Gilbertson L. N., "Optimization of Properties of Ti-15Mo-2.8Nb-0.2Si & Ti-15M0-2.8Nb-0.2Si-.26O Beta Titanium Alloy for Application in Prosthetic Implant," *Medical Applications of Titanium and its Alloys: the Material and Biological Issues, ASTM STP 1272*, S. A. Brown and J. E. Lemons Eds., ASTM International, 1996, pp. 88–95.

Journal of ASTM International, June 2005, Vol. 2, No. 6
Paper ID JAI12818
Available online at www.astm.org

M. Niinomi,[1] *T. Akahori,*[2] *Y. Hattori,*[3] *K. Morikaw,*[4] *T. Kasuga,*[5] *H. Fukui,*[6] *A. Suzuki,*[7] *K. Kyo*[8] *and S. Niwa*[9]

Super Elastic Functional β Titanium Alloy with Low Young's Modulus for Biomedical Applications

ABSTRACT: The low modulus β type titanium alloy, Ti-29Nb-13Ta-4.6Zr, was designed, and then the practical level ingot of the alloy was successfully fabricated by Levicast method. The mechanical and biological compatibilities, and super elastic behavior of the alloys were investigated in this study. The mechanical performance of tensile properties and fatigue strength of the alloy are equal to or greater than those of conventional biomedical Ti-6Al-4V ELI. Young's modulus of the alloy is much lower than that of Ti-6Al-4V ELI, and increases with the precipitation of α phase or ω phase in the β matrix phase. The compatibility of the alloy with bone is excellent. Low modulus of the alloy is effective to enhance the healing of bone fracture and remodeling of bone. Super elastic behavior is observed in Ti-29Nb-13Ta-4.6Zr conducted with short time solution treatment after heavy cold working. Total elastic strain in that case is around 2.8 %. The mechanism of the super elastic behavior of Ti-29Nb-13Ta-4.6Zr is still unclear.

KEYWORDS: Ti-29Nb-13Ta-4.6Zr, β–type titanium alloy, low modulus, biomedical application, biocompatibility, super elastic behavior

Introduction

Pure titanium and α + β type Ti-6Al-4V ELI alloys are currently used widely as structural biomaterials for instruments for replacing failed hard tissues such as artificial hip joints, dental implants, etc., because they have excellent specific strength and corrosion resistance, no allergic problems and the greatest biocompatibility among the metallic biomaterials. They occupy almost the whole market of titanium biomaterials. However, other new titanium alloys for biomedical applications have been registered with ASTM standardizations after pure titanium and Ti-6Al-

Manuscript received 14 June 2004; accepted for publication 24 November 2004; published June 2005. Presented at ASTM Symposium on Titanium, Niobium, Zirconium, and Tantalum for Medical and Surgical Applications on 9-10 November 2004 in Washington, DC.

[1] Professor, Department of Production Systems Engineering, Toyohashi University of Technology, 1-1, Hibarigaoka, Tempaku-cho, Toyohashi 441-8580, Japan.

[2] Research Associate, Department of Production Systems Engineering, Toyohashi University of Technology, 1-1, Hibarigaoka, Tempaku-cho, Toyohashi 441-8580, Japan.

[3] Department of Materials Science and Engineering, Faculty of Science and Technology, Meijo University Aichi, 101 Shiogamaguchi , Tempaku-ku, Nagoya, 468-8502, Japan.

[4] Research Assistant, Department of Orthopedic Surgery, Aichi Medical University, Nagakute, Aichi, 480-1195, Japan.

[5] Professor, Department of Materials Science, Nagoya Institute of Technology, Gokiso-cho, Showa-ku, Nagoya 466-8555, Japan.

[6] Professor, Department of Dental Materials Science, School of Dentistry, Aichi-Gakuin University, 1-100, Kusumoto-cho, Chikusa-ku, Nagoya 464-8650, Japan.

[7] Vice Manager, R & D Laboratory, Daido Steel Co., Ltd., 2-30, Daido-cho, Minami-ku, Nagoya 457-8545, Japan.

[8] Technical Scientist, Surface Characterization Division, The Institute of Physical and Chemical Research, 2-1, Hirosawa, Wako-shi, Saitama 351-0198, Japan.

[9] Professor Emeritus, Aichi Medical University, Nagakute, Aichi, 480-1195, Japan.

4V ELI, including Ti-15Mo etc. [1].

β type titanium alloys have been developed or are being developed in order to obtain low modulus titanium alloys, because the low modulus is said to be effective to enhance the bone healing and remodeling. The moduli of α + β type titanium alloys are still much greater than those of cortical bone, although the modulus of titanium alloys is much smaller than that of Co-Cr type alloys and SUS stainless steels used for biomedical applications [2]. The recent trend in research and development of titanium alloys for biomedical applications is to develop the low modulus β type titanium alloys composed of nontoxic and nonallergic elements with excellent mechanical properties and workability [3].

In addition to low modulus, functions like shape memory, super elasticity, etc., are required for metallic biomaterials. Research and development on shape memory and super elastic beta type titanium alloys for biomedical applications are energetically done. Many types of beta titanium alloys composed of nontoxic and nonallergic elements, which have potential to achieve excellent shape memory effect or super elasticity are proposed [4]. In general, super elastic behavior is related with stress or strain induced martensitic transformation and its reverse transformation [5]. Ti-Nb-Ta-Zr system beta type titanium alloy made by powder metallurgy for biomedical applications shows super elastic behavior when it is conducted with some thermomechanical treatment. However, it seems not to be related with deformation induced transformation and its reverse transformation, but with some unknown mechanism [6].

A low modulus β type titanium alloy composed of nontoxic and nonallergic elements with excellent mechanical properties and workability for biomedical applications was designed in this study. The mechanical biocompatibility and biological compatibility of the designed alloy were investigated. The themomechanical treatment to induce super elasticity was investigated.

Experimental Procedures

Alloying Elements and Alloy Composition

Nontoxic elements were selected based on the data of cyto-toxicity of pure metals [7] and the data of polarization resistance and biocompatibility of representative metallic biomaterials and pure metals [8]. Finally, Nb, Ta, and Zr were selected as alloying elements for Ti. Therefore, Ti-Nb-Ta-Zr system β-type titanium alloys with low modulus were determined to be developed in this study. The compositions of the candidate alloys were determined using the d-electron alloy design method developed by Morinaga et al. [9]. Finally, the most promising alloy for the practical use for biomedical applications was found to be Ti-29Nb-13Ta-4.6Zr based on the balance of strength and ductility balance obtained from tensile tests on the specimens fabricated from the laboratory size ingot (around 45 g) made by tri-arc furnace.

Melting and Processing

The practical level ingot of Ti-29Nb-13Ta-4.6Zr (hereafter TNTZ) around 20 kg was fabricated by the levitation casting (Levicast) method [10]. The ingot was first forged at 1223 K and then forged at 1123 K to finish the bars with a diameter of 20 mm or 12 mm.

Material Preparation

The forged bars with a diameter of 20 mm were cold rolled after solution treatment at 1063 K for 3.6 ks by reductions of 87.5 % to plates with a thickness of 2.5 mm at room temperature in air. The rolled plates were conducted with solution treatment at 1063 K for 3.6 ks followed by water quenching (ST), and then aging (STA) at various temperatures. Some plates were directly aged at various temperatures after cold rolling. The solutionized forged bar was also cold rolled by various reductions between 50 and 95 %. They were then solution treated at 1073 K for various times between 0.01 ks and 1.8 ks.

On the other hand, the forged bars with a diameter of 12 mm were conducted with solution treatment at 1063 K for 3.6 ks followed by water quenching, machined to 10 mm diameter bars and then cold swaged to bars with various diameters up to 4 mm according to the reduction ratio. Cold swaged bar with a diameter of 4 mm was also extruded to the wire with diameters of 1 mm and 0.3 mm. In this case, annealing was carried out at a certain stage during wire drawing.

Mechanical Testing

Tensile and fatigue test specimens with a width of 1.2 mm, a thickness of 1.5 mm, and a cross-sectional area of 4.5 mm^2, and specimens for measuring Young's modulus with a length of 5.6 mm, a width of 1.2 mm, and a thickness of 1.5 mm were machined from the heat treated plates so the tensile axis was parallel to the rolling direction. The tensile test specimens with heat treatments were wet polished using waterproof emery papers up to #1500 and then buff polished. The specimens for measuring Young's modulus were wet polished using waterproof emery papers up to #1500.

For cold swaged bars with a diameter of over 4 mm, tensile specimens with a gage diameter of 6.35 mm and a gage length of 25.4 were machined so the tensile axis was parallel to the swaging direction. For the cold swaged bar with a diameter of 4 mm, tensile specimens with a gage diameter of 2.4 mm and a gage length of 9.6 mm were machined so the tensile axis was parallel to the swaging direction. In this case, the surface condition of the specimen was as machined.

Tensile tests were carried out using an Instron-type machine at a crosshead speed of 8.33 $\times$ 10^{-6} m/s in air at room temperature. Tensile loading-unloading tests were carried out using an Instron-type machine. The tensile loading-unloading was continuously done with every 0.5 % increasing elastic strain up to a total elastic strain of 4.0 %.

Fatigue tests were carried out using an electro-servo-hydraulic machine at a frequency of 10 Hz with a stress ratio, $R = 0.1$, under the tension-tension mode in air at room temperature in air at 295 K and Ringer's solution at 310 K. Young's modulus was measured using a resonance method in air at room temperature. For comparison, Young's modulus of Ti-6Al-4V ELI was also measured.

Microstructual Analysis

The constituted phases of each heat-treated alloy were examined on the chuck part of the tensile specimen through an X-ray diffraction analysis and observations using a transmission electron microscopy (TEM). X-ray analysis was carried out using a Cu target with an accelerating voltage of 40 kV and a current of 30 mA. TEM observations were carried out with an acceleration voltage of 200 kV.

Evaluation of Biological Compatibility

Columnar specimens with a size of ϕ5 mm × 10 mm were machined from the cold swaged bar of Ti-29Nb-13Ta-4.6Zr with a diameter of 5 mm. In this case, the same size columnar specimens of commercial Ti-6Al-4V ELI and SUS 316 stainless steel for biomedical applications were also prepared. The specimens were implanted into lateral femoral condyles of Japanese white rabbits weighing 2.5–3.0 kg under intravenous anesthesia. Then, histological observation was performed with C. M. R. (Contact Micro Radiogram) at 4 and 8 weeks after the implantation

Evaluation of Mechanical Biocompatibility

Experimental tibia fractures were made by oscillating saw just below the tibial tuberosity of Japanese white rabbits weighing 2.5–3.0 kg under intravenous, and intramedullary fixations were performed by the rods of Ti-29Nb-13Ta-4.6Zr cut from the cold swaged bars, and commercial Ti-6Al-4V ELI and SUS 316 L stainless steel. The size of each rod was ϕ3 mm × 60 mm. In order to observe continuously the state of fracture healing, X-ray pictures were taken at every 2 weeks for 22 weeks.

Results and Discussion

Distribution of Alloying Elements in Ingot

The alloy contains the elements of Ta and Nb that have greater specific gravity and higher melting points as compared with those of Ti.

The distribution of each element was checked. The distribution of each element from the top through the bottom of the ingot is shown in Fig. 1. It is clear that each element distributes homogeneously from the top to bottom of the ingot, and is in the range for target content value.

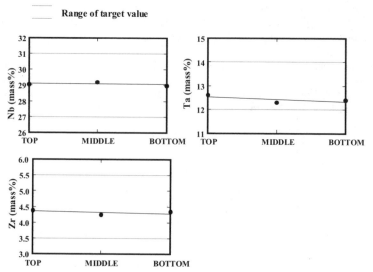

FIG. 1—*Distribution of Nb, Ta, or Zr from top through bottom of ingot of Ti-29Nb-13Ta-4.6Zr.*

Tensile Properties

The balance of tensile strength and elongation of TNTZ conducted with various aging after solution treatment or directly after cold rolling is shown in Fig. 2 with that of annealed Ti-6Al-4V ELI [11] [12]. The strength and elongation of TNTZ can be controlled variously by conducting heat treatment or thermomechanical treatment, and is better than that of Ti-6Al-4V ELI by conducting proper heat treatment or thermomechanical treatment.

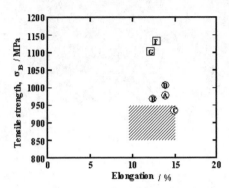

FIG. 2—*Tensile strength of Ti-6Al-4V ELI conducted with various aging treatments after solution treatment, and Ti-29Nb-13Ta-4.6Zr conducted with various aging treatments after solution treatment or after cold rolling;* ▨: *Ti-6Al-4V ELI, Ti-29Nb-13Ta-4.6Zr (A; (1033 K, 1.8 ks+673 K, 259.2 ks), B: (1033 K, 1.8 ks + 598 K, 100.8 ks), C:(1033 K, 1.8 ks + 723 K, 259.2 ks), D: (1063 K, 1.8 ks + 673 K, 259.2 ks), F: (Cold rolling +723 K, 100.8 ks), G: (Cold rolling + 723 K, 259.8 ks).*

Fatigue Strength

Stress-fatigue life (the number of cycles to failure) curves, that is, S-N curves, obtained from plain fatigue tests on TNTZ conducted with solution treatment (ST), $TNTZ_{ST}$, and aging treatments after solution treatment (STA) are shown in Fig. 3. The ranges of fatigue strength of conventional Ti-6Al-4V ELI and Ti-6Al-7Nb for biomedical applications are quoted from literatures [13–15] and are also shown in the same figure as a comparison. The maximum stress, at which the specimen survives 10^7 cycles, is defined as a fatigue limit in this study.

The fatigue strength of ST alloy is the lowest in both the low cycle fatigue life region, where the number of cycles to failure is less than 10^4 cycles, and the high cycle fatigue life region, where the number of cycles to failure is over 10^5 cycles. The fatigue limit of ST alloy is approximately 320 MPa.

On the other hand, the fatigue strength of each STA alloy is similar to that of Ti-6Al-4V ELI and Ti-6Al-7Nb, and is around two times greater than that of $TNTZ_{ST}$. The fatigue limit of $TNTZ_{ST}$ aged at 673 K is around 700 MPa, which is the greatest and nearly upper limit of fatigue limit of Ti-6Al-4V ELI.

S-N curves obtained from corrosion fatigue tests on $TNTZ_{ST}$ and $TNTZ_{ST}$ aged at 673 K in Ringer's are shown in Fig. 4 with those of $TNTZ_{ST}$ and $TNTZ_{ST}$ aged at 673 K in air shown in Fig. 4. The fatigue strength in Ringer's solution is equal to that in air for both as-solutionized and aged conditions. Therefore, the fatigue strength of TNTZ is not degraded in Ringer's solution.

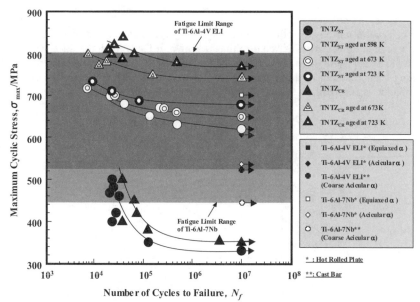

FIG. 3—*S-N curves of Ti-29Nb-13Ta-4.6 in as-solutionized conditions   (TNTZ$_{ST}$) and as-cold rolled conditions (TNTZ$_{CR}$), and TNTZ$_{ST}$ and TNTZ$_{CR}$ conducted with aging at 598 K, 673 K and 723 K for 259.2 ks with fatigue limit ranges of Ti-6Al-4V ELI and Ti-6Al-7Nb.*

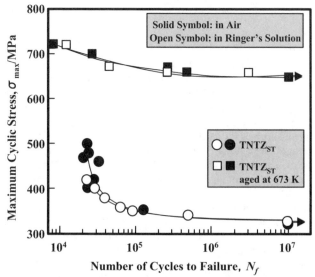

FIG. 4—*S-N curves of Ti-29Nb-13Ta-4.6Zr in as-solutionized conditions (TNTZ$_{ST}$) and TNTZ$_{ST}$ aged at 673 K for 259.2 ks in air and Ringer's solution.*

Young's Modulus

Young's moduli of TNTZ and Ti-6Al-4V ELI are shown in Fig. 5. For TNTZ, Young's moduli in various aging conditions are also shown in Fig. 5. Young's modulus of TNTZ is much smaller than that of Ti-6Al-4V ELI, even in the case of in aged conditions. In the case of the smallest Young's modulus, around 65 GPa is obtained in as-solutionized conditions (ST). This value of Young's modulus is nearly the half of that of Ti-6Al-4V ELI. Furthermore, Young's modulus of TNTZ is changed by aging treatment because α or ω phase with greater Young's modulus comparing with β phase precipitates in β matrix phase, as will be mentioned in the section of microstructure. Therefore, Young's modulus of TNTZ can be controlled by aging treatment.

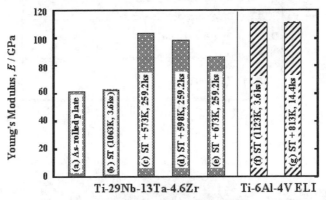

FIG. 5—*Young's moduli of Ti-29Nb-13Ta-4.6Zr in as-rolled, as-solutionized (ST) and aged conditions at various temperatures, and Ti-6Al-4V ELI in as-solutionized (ST) and aged conditions.*

Microstructure

Microstructure of TNTZ conducted with solution treatment showed only β phase with an average diameter of 20 μm. In TNTZ aged at a temperature between 573 and 673 K after solution treatment, it was difficult to observe precipitated phases by light microscopy because they were too fine to identify. Therefore, X-ray analysis and TEM observations were carried out to identify the phases in aged alloys.

X-ray diffraction profiles of the alloy conducted with solution treatment, and aging treatments are shown in Fig. 6, revealing that α phase or α and ω phases are precipitated in TNTZ aged at a temperature between 573 and 673 K after solution treatment.

Figure 7 shows TEM micrograph and diffraction pattern of TNTZ aged at 673 K for 259.2 ks after solution treatment. Fine plate like α phases with two variants are precipitated homogeneously in β phase. The α phases with two variants have a relation of crystal orientation rotated around 90° to β [100] phase. In addition, the precipitation of ω phase is identified using the TEM diffraction pattern as shown in Fig. 8.

Therefore, the increasing strength and fatigue strength of the alloy result from the precipitation of α phases or α and ω phases in the β matrix phase.

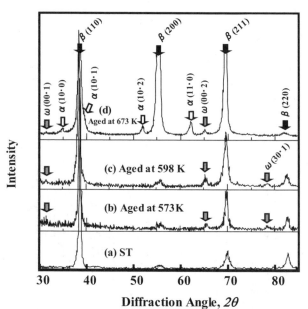

Diffraction Angle, 2θ

FIG. 6—*X-ray diffraction profile of Ti-29Nb-13Ta-4.6Zr conducted with* (a) *solution treatment (ST), or aged for 259.2 ks at* (b) *573 K,* (c) *598 K or* (d) *673 K after solution treatment.*

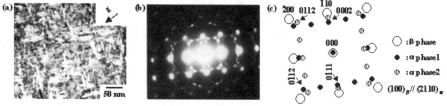

FIG. 7—*Transmission electron micrograph, diffraction pattern and key diagram of Ti-29Nb-13Ta-4.6Zr aged at 673K for 259.2ks after solution treatment at 1063K for 3.6ks:* (a) *dark field image of a phase,* (b) *diffraction pattern, and* (c) *key diagram.*

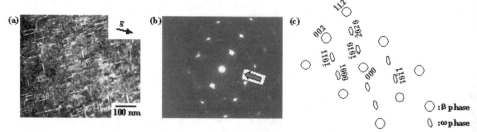

FIG. 8—*Transmission electron micrograph, diffraction pattern and key diagram of Ti-29Nb-13Ta-4.6Zr aged at 673K for 259.2ks after solution treatment at 1063K for 3.6ks. Arrow shows w phase spot:* (a) *dark field image of w phase,* (b) *diffraction pattern, and* (c) *key diagram.*

Tensile Properties of Cold Swaged Alloy

Tensile properties of cold swaged TNTZ conducted with solution treatment are shown in Fig. 9 as a function of cold work ratio. Tensile strength, σ_B, and 0.2 % proof stress, $\sigma_{0.2}$, increase with increasing cold work ratio. The strength of the alloy at the greatest cold work ratio reaches a similar value of conventional Ti-6Al-4V ELI. Elongation and reduction of area decrease at about 20 % cold work ratio, and are then almost constant with cold work ratio. For example, the elongation at the greatest cold work ratio is over 15 %, which is a relatively greater value of elongation. Young's modulus is almost constant with increasing cold work ratio. Therefore, the strength of TNTZ can be increased up to the equivalent strength value of conventional Ti-6Al-4V ELI with keeping Young's modulus low constant value by cold working.

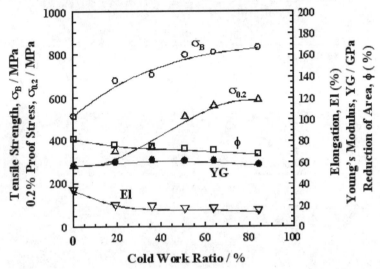

FIG. 9—*Relationship between tensile properties and cold work ratio of cold swaged bar of Ti-29Nb-13Ta-4.6Zr.*

Biological Compatibility

Biocompatibility with Bone—The contact micro radiogram (C. M. R.) of the boundaries of bone and TNTZ (in this case, TNTZ$_{ST}$), Ti-6Al-4V, or SUS 316 L stainless steel implanted into lateral femoral condyles of the rabbit is shown in Fig. 10. Each specimen is surrounded by newly formed bone, and the bone tissue shows direct contact partially with specimen. However, the extent of the direct contact is greater in TNTZ as compared with Ti-6Al-4V and SUS 316 L stainless steel. Therefore, the biocompatibility of TNTZ with bone is excellent.

Effect of Low Modulus—In order to confirm the advantage of low modulus for bone healing and remodeling, using rabbits, experimental tibial fracture was made by oscillating saw at just below the tibial tuberosity, and intramedullary rod made of low modulus TNTZ, Ti-6Al-4V ELI, or SUS 316 L stainless steel was driven into the intramedullary canal to fix the fracture. The observation of the state of bone healing, remodeling, and atrophy was continuously done with taking an X-ray picture every 2 weeks up to 24 weeks. The results are shown in Fig. 11.

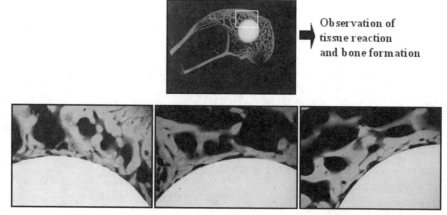

Ti-29Nb-13Ta-4.6Zr   Ti-6Al-4V ELI   SUS 316L stainless steel

FIG. 10—*C. M. R. photograph of boundary of each specimen and bone at 8 weeks after implantation.*

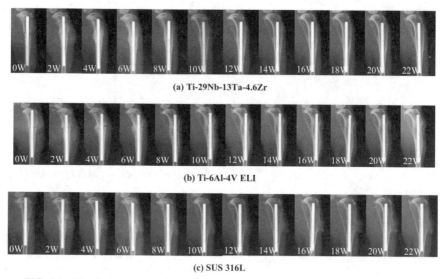

(a) Ti-29Nb-13Ta-4.6Zr

(b) Ti-6Al-4V ELI

(c) SUS 316L

FIG. 11—*Healing process of bone fracture (A) from 0 to 22 weeks after surgery.*

The outline of fracture callus is very smooth with bone remodeling in TNTZ. Similar phenomenon is observed at 8 weeks in Ti-6Al-4V ELI and SUS 316L. In TNTZ, the amount of the fracture callus is relatively small, and gradually decreases from 6 weeks, and then there are no traces of fracture at 10 weeks after the fixation. After 10 weeks, no changes can be observed up to 18 weeks. However, a little atrophic change is observed at the posterior tibial bone after 20 weeks. In Ti-6Al-4V ELI, the callus formation and the bone remodeling are almost similar to those in TNTZ, but slower as compared with TNTZ. A little atrophic change is observed at 18 weeks. In SUS 316 L stainless steel, a large amount of the fracture callus is observed, and remains up to the end of the follow-up period. Bone atrophy seems to be occurring at the posterior proximal tibial bone at 10 weeks, and becomes obvious every 2 weeks. The posterior tibial bone comes to be very thin at 24 weeks.

The cross section of tibia implanted with each rod at 24 weeks after implantation is shown in Fig. 12. The microstructure of bone formed around the rod made of TNTZ or Ti-6Al-4V ELI shows a number of osteons, which is the result of internal remodeling of cortical bone, but in the case of SUS 316L stainless steel the bone structure shows lamination of cortical bone with the absence of osteons. The absence of osteons suggests the bone strophy in SUS 316L stainless steel.

Therefore, low modulus titanium alloy, TNTZ, is found to improve the load transmission issue of the current metal implants with the high modulus.

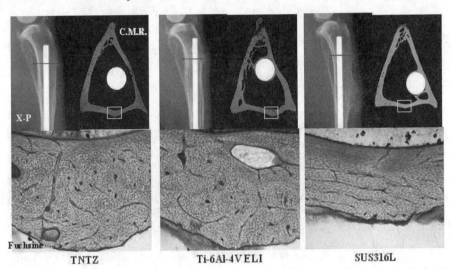

FIG. 12—*Cross section of tibia at 24 weeks after implantation.*

Super Elasticity

Super Elastic Behavior of Wire—The tensile loading-unloading curve of TNTZ wires with diameters of 1.0 and 0.3 mm are shown in Fig. 13. In both cases, relationship between loading stress and strain curve shows elastic-plastic behavior from the very early stage of deformation. By unloading, the elastic back strain is very large. The maximum elastic strain is 2.8 % for the wire with a diameter of 1.0 mm and 2.9 % for the wire with a diameter of 0.3 mm. As a result of these elongations, we can conclude that these wires show super elastic behaviors. In general, super elastic behavior has been attributed to deformation induced martensitic transformation and its reverse transformation. Therefore, DSC analysis was done on these wires in the temperature range from 223 K through 423 K. In this temperature range, the peak concerning martensitic transformation could not be detected. The X-ray analysis was also done on the deformed wire with a diameter of 1 mm. The super elastic behavior of the wires of TNTZ was, therefore, considered not to be related to deformation induced transformation and its reverse transformation.

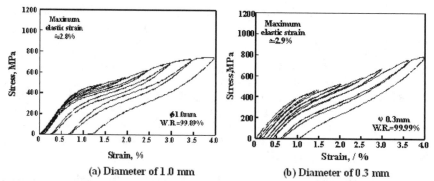

FIG. 13—*Tensile loading-unloading curves of extruded wires of Ti-29Nb-13Ta-4.6Zr with diameters of (a) 1.0 mm and (b) 0.3 mm, respectively.*

Super Elastic Behavior of Rolled Plates—Figure 14 shows the tensile loading-unloading stress-strain curves of TNTZ conducted with cold swaging by various reduction ratios from 0 % through 84 % after solution treatment. Every tensile loading-unloading curve is very similar to that observed in general metallic materials, and total elastic strain is very small. Figure 15 shows the tensile loading-unloading stress-strain curves of TNTZ conducted with cold rolling by reduction ratios of 0, 50, and 90 % after solution treatment. In this case, every tensile loading-unloading curve is also very similar to that observed in general metallic materials, and total elastic strain is very small. Figure 16 shows tensile loading-unloading stress-strain curves on TNTZ conducted with solution treatment at 1073 K for various times after cold rolling by a reduction ratio of 95 %. Every tensile loading-unloading stress-strain curve shows super elastic behavior, and large total elastic strain. However, a solution treatment time of 0.3 ks results in the largest total strain of 2.8 %. This value is nearly the same as that obtained in wires of TNTZ stated above.

X-ray diffraction patterns of TNTZ conducted with solution treatment at 1073 K for 0.3 ks after cold rolling by a reduction ratio of 95 % before and after tensile loading-unloading test are shown in Fig. 17. The peak of martensite α" is observed before tensile loading-unloading test, although this peak was not observed in a solutionized TNTZ. The peak of α" is also observed after tensile loading-unloading test, but the intensity of the peak is not changed as compared to that observed before tensile loading-unloading test. In this case, the existence of α" is the key factor for creating super elasticity. However, the mechanism of the super elastic behavior in this study is still unclear. Further investigation is, therefore, needed for achieving a better understanding of the mechanism of the super elastic behavior observed in TNTZ.

Conclusions

Low modulus β type titanium alloy composed of nontoxic and nonallergic elements of Nb, Ta and Zr, Ti-29Nb-13Ta-4.6Zr, for biomedical applications was designed for this study. The basic mechanical biocompatibility and biological biocompatibility of practical level Ti-29Nb-13Ta-4.6Zr were investigated. Finally, the super elastic behavior of Ti-29Nb-13Ta-4.6Zr conducted with severe cold working and solution treatment was examined. The following results were obtained.

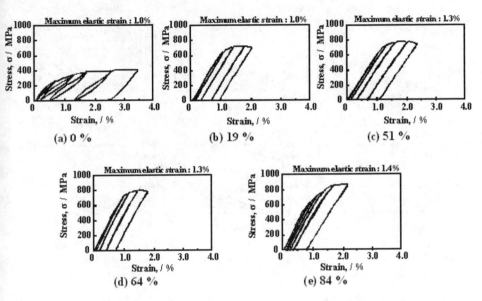

FIG. 14—*Tensile loading-unloading stress-strain curves of cold swaged TNTZ by various work ratios from 0–84 %.*

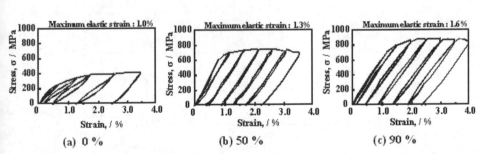

FIG. 15—*Tensile loading-unloading stress-strain curves of cold rolled TNTZ with various work ratios form 0–90 %.*

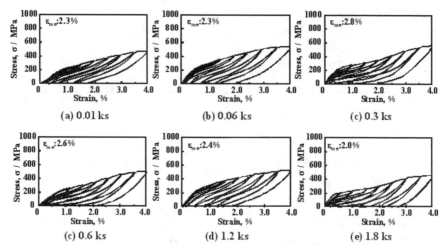

FIG. 16—*Tensile loading-unloading stress-strain curves of TNTZ conducted with solution treatment for various times from 0.01–1.8 ks after cold rolling by a reduction ratio of 95 %.*

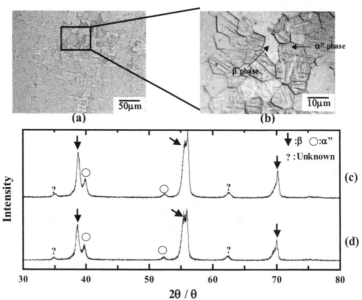

FIG. 17—*Optical micrographs and X-ray diffraction patterns of TNTZ conducted solution treatment at 1073 K for 0.3 ks after cold rolling by a reduction ration of 95 %.*

a.   *Optical micrograph (low magnification)*
b.   *Optical micrograph (high magnification)*
c.   *Before tensile loading-unloading test*
d.   *After tensile loading-unloading test*

1. The practical level ingot of Ti-29Nb-13Ta-4.6Zr is successfully fabricated in this study.
2. The balance of strength and ductility of Ti-29Nb-13Ta-4.6Zr is equivalent to or better than those of Ti-6Al-4V ELI by aging treatment.
3. The fatigue strength of Ti-29Nb-13Ta-4.6Zr is equivalent to that of Ti-6Al-4V ELI.
4. Young's modulus of Ti-29Nb-13Ta-4.6Zr is much smaller than that of Ti-6Al-4V ELI.
5. ω phase precipitates or both ω and α phases precipitates in β phase in Ti-29Nb-13Ta-4.6Zr conducted with aging after solution treatment at a temperature between 573 K and 673 K.
6. The strength of Ti-29Nb-13Ta-4.6Zr can be increased with keeping Young's modulus low constant value by cold working.
7. The biocompatibility of Ti-29Nb-13Ta-4.6Zr with bone is better than that of SUS 316 stainless steel or Ti-6Al-4V ELI.
8. The low modulus Ti-29Nb-13Ta-4.6Zr enhances the healing of bone fracture and remodeling of bone.
9. The cold extruded wire of Ti-29Nb-13Ta-4.6Zr shows super elastic behavior.
10. Ti-29Nb-13Ta-4.6Zr conducted with severe cold rolling followed by short-term solution treatment also shows super elastic behavior.
11. The mechanism of super elastic behavior of Ti-29Nb-13Ta-4.6Zr seems not to be related with deformation induced martensite transformation and its reverse transformation.

References

[1] ASTM Standard F 2066-01, 2001: Specification for Wrought Titanium-15 Molybdenum Alloy for Surgical Implant Applications, ASTM International, West Conshohocken, PA, pp. 1605–1608.
[2] Niinomi, M., 2001, "Recent Metallic Materials for Biomedical Applications," *Metall. Mater. Trans. A*, Vol. 32A, No. 12, pp. 477–486.
[3] Kuroda, D., Niinomi, M., Morinaga, M., Kato, Y., and Yashiro, T., 1998, "Design and Mechanical Properties of New Beta Type Titanium Alloys for Implant Materials," *Materials Science and Engineering A*, Vol. A243, pp. 244–249.
[4] Niinomi, M, 2003, "Recent Research and Development in Titanium Alloys for Biomedical Applications and Healthcare Goods," *Science and Technology for Advanced Materials*, Vol. 4, pp. 445–454.
[5] Akahori, T., Niinomi, M., Higuchi, T., and Morii, K., 2003, "Microstructure and Mechanical Properties of Ti-Ni and Ti-Ni-Co Type Shape Memory Alloys," *JIM*, Vol. 67, No. 10, pp. 528–536.
[6] Sakaguchi, N., Niinomi, M., and Akahori, T., 2004, "Deformation Behaviors of Ti-Nb-Ta-Zr System Alloys for Biomedical Applications," *Mat. Transactions*, Vol. 45, No. 5.
[7] Kawahara, H., Ochi, S., Tanetani, K., Kato, K., Isogai, M., Mizuno, Y., Yamamoto, H., and Yamaguchi, A., 1963, "Biological Test of Dental Materials, Effect of Pure Metals upon the Mouse Subcutaneous Fibroblast, Strain L-cells in Tissue Culture," *J. Jpn,. Soc. Dent. Apparat. & Mater.*, Vol. 4, pp. 65–75.
[8] Steinemann, S. G., 1980, "Corrosion of Surgical Implants-in Vivo and in Vitro Test," *Evaluation of Biomaterials*, G. D. Winter, J. L. Leray, and K. de Groot, Eds., John Wiley & Sons, New York, pp. 1–34.
[9] Morinaga, M., Kato, M., Kamimura, T., Fukumoto, M., Harada, I., and Kubo, K., 1993, "Theoretical Design of β-type Titanium Alloys," *Proceedings, Titanium'92: Science and Technology*, Vol. 1, F. H. Froes, Ed., TMS, Warrendale, PA, pp. 217–224.

[10] Demukai, N., 2001, "Development of a New Type Titanium Casting Technology/ LEVICAST Process," *Proceedings, PRICM4*, Japan Institute of Metals, pp. 369–372.

[12] Niinomi, M., 1998, "Mechanical Properties of Biomedical Titanium Alloys," *Materials Science and Engineering A*, Vol. A243, pp. 231–236.

[13] Akahori, T. and Niinomi, M., 1998, "Fracture Characteristics of Fatigued Ti-6Al-4V ELI as an Implant Materials," *Materials Science and Engineering A*, Vol. A243, T. Akahori and M. Niinomi., Eds., pp. 237–243.

[14] Akahori, T., Niinomi, M., and Suzuki, A., 2001, "Improvement in Mechanical Properties of Dental Cast Ti-6Al-7Nb by Thermochemical Processing," *Metallurgical and Materials Transactions A*, Vol. 32A, No. 12, pp. 503–510.

[15] Akahori, T., Niinomi, M., and Fukunaga, K., 2000, "An Investigation of the Effect of Fatigue Deformation on the Residual Mechanical Properties of Ti-6Al-4V ELI," *Metall. Mater. Trans. A*, Vol. 31A, pp. 1049–1948.

[16] Akahiri, T., Niinomi, M., Fukunaga, K., and Inagaki, I., 2000, "Small Fatigue Crack Initiation and Propagation Characteristics of Biomedical Ti-6Al-7Nb with Relating Microstructure," *Metall. Mater. Trans. A*, Vol. 31A, pp. 1949–1958.

Journal of ASTM International, October 2005, Vol. 2, No. 9
Paper ID JAI13084
Available online at www.astm.org

Donald W. Petersen,[1] *Jack E. Lemons, Ph.D., and Linda C. Lucas, Ph.D.*

Comparative Evaluations of Surface Characteristics of cp Titanium, Ti-6Al-4V and Ti-15Mo-2.8Nb-0.2Si (Timetal® 21SRx)

ABSTRACT: Commercially pure titanium (cpTi), Ti-6Al-4V (Ti64), and Ti-15Mo-2.8Nb-0.2Si (21SRx), with three unique atomic, alpha, alpha-beta, and beta grain structures, respectively, were subjected to three different surface treatments: cleaning, nitric acid passivation, and heat treatment. Experiments were conducted to determine the effects of the type of material and surface modifications on the substrate microstructure, surface oxide composition and thickness, and resultant corrosion behavior.

Metallography showed the cpTi groups were an equiaxed single alpha phase material, the Ti64 groups a dual-phase alpha-beta material, and the 21SRx groups an equiaxed beta material. The different surface treatments did not alter the substrate microstructures of any groups.

Spectroscopic (AES) results showed typical titanium and titanium alloy spectra with dominant Ti and O peaks for all sample groups, indicative of titanium dioxide. In addition, small Al and Mo peaks were detected throughout the surface oxides of the Ti64 and 21SRx specimens, respectively. AES depth profiling showed no significant difference in the oxide thickness between all the Cleaned and Passivated groups regardless of metal or alloy group. However, all the Heat Treated groups had significantly thicker oxides.

In general, corrosion results showed Passivated and Heat Treated groups to have similar corrosion properties and significantly improved corrosion resistances compared to the Cleaned groups. All impedance spectra fit into the Randles equivalent circuit model, and all sample groups exhibited near ideal capacitive behavior ($\phi \cong 90°$) expected for titanium and its alloys.

KEYWORDS: titanium, titanium alloy, beta-titanium, corrosion properties, surface oxide, and surface treatments

Introduction

Titanium and titanium alloys, such as Ti-6Al-4V, are the most commonly used metallic implant biomaterials. Biocompatibility has often been attributed to excellent corrosion properties [1–3]. In addition, titanium and its alloys have relatively high strength and low elastic modulus. A lower implant elastic modulus has been proposed to more evenly disperse mechanical loading throughout the implant interface, which in turn minimizes "stress shielding" of the host bone, resulting in less bone atrophy [4,5].

The excellent corrosion and biocompatibility properties of titanium and its alloys are attributed to the formation of a stable, protective oxide film on their surfaces [6]. The composition and thickness of surface oxides are known to depend on the substrate and the environment in which the oxide is formed [7–9]. Oxides that form on mechanically polished or machined specimens are generally classified as native oxides. These oxides form quickly on

Manuscript received 7 October 2004; accepted for publication 27 January 2005; published October 2005.
Presented at ASTM Symposium on Titanium, Niobium, Zirconium, and Tantalum for Medical and Surgical Applications on 9-10 November 2004 in Washington, DC.
[1] University of Alabama at Birmingham, Birmingham, AL 35264.

freshly polished or machined specimens on exposure to air or other environments that contain oxygen (e.g., water or tissue fluids). Native oxides are thin (less than 10 nm thick), amorphous in structure, and composed primarily of titanium and oxygen [7,10]. However, in the case of alloys, alloy elements such as Al and V have been found within the oxide coating [10–13].

Materials that are exposed to stronger oxidizing solutions than air or water are said to have undergone passivation. In the case of titanium and its alloys, a commonly used method for passivation is by using a nitric acid solution in accordance to the protocol specified in Standard Practice for Surface Preparation and Marking of Metallic Surgical Implants (ASTM F 86-01). Titanium and titanium alloy specimens that have been passivated using this protocol form thin oxides, much like the native oxides, with thicknesses of less than 10 nm [11,14–17]. The details of the effect of passivation on corrosion properties remain as a topic of research. Generally, improved corrosion properties are believed to result from this process; however, conflicting results have been reported [11,18].

Heat treatment at elevated temperatures in air can also further oxidize titanium and its alloys. Heat treatments have been shown to increase the thickness and crystallinity of the oxide, with the extent of such increases depending on the temperature and duration of the heat treatment [19,20]. It is generally thought that thicker oxides, which result from oxidation at elevated temperatures, may improve corrosion properties [21–25].

In a number of different implant systems used in medicine and dentistry, surface treatments have included cleaning, acid passivation, and heat treatment at elevated temperatures. Therefore, on a relative basis one could question what specific treatment offers the optimum biocompatibility characteristics.

It is the goal of this study to compare the effects of nitric acid solution passivation and heat treatment on the resultant surface oxide and corrosion properties of titanium and titanium alloys with different chemical compositions and microstructures. In this study, three materials were included: one with a single-phase alpha (hcp) microstructure, cpTi; one with a mixed alpha-beta (hcp and bcc) microstructure, Ti64; and one with a single-phase beta (bcc) microstructure, 21SRx. The microstructure of each material was documented using standard metallographic techniques. The chemistry and thickness of the surface oxides for each material in three different conditions were determined using Auger electron spectroscopy (AES), and the corrosion properties were determined using DC polarization tests and AC electrochemical impedance spectroscopy (EIS).

Materials and Methods

Materials

Discs, 15 mm in diameter and 3 mm thick, were cut from bar stock from the titanium metals and alloys listed in Table 1 using a diamond saw. The discs were subsequently mounted in polymethylmethacrylate (PMMA) and wet ground with silicon carbide paper to a 600-grit finish, and then polished using alumina powder slurry to a 0.3 um finish.

TABLE 1—*List of titanium and titanium alloys used in this study.*

Metal or Alloy	ASTM Specification	Abbreviation
cp Titanium	F-67 grade 4	CpTi
Ti-6Al-4V	F-136	Ti64
Ti-15Mo-2.8Nb-0.2Si	....	21SRx

Surface Treatments

The discs from each of the three metal groups were then divided into three experimental sample groups, which are listed in Table 2.

TABLE 2—*Surface treatments used in this study.*

Surface Treatment		Description
Cleaned	(C)	Cleaned ultrasonically in a series of solvents
Passivated	(P)	C + passivated in 40 % nitric acid for 30 min
Heat Treated	(HT)	C + P + heat treated at 350°C in air for 1 h

Cleaned discs (C) were obtained by ultrasonically cleaning the discs for 10 min in each of the following solutions: distilled water, benzene, acetone, and ethanol. The discs were thoroughly rinsed with distilled water between solvents. Passivated discs (P) were obtained by cleaning discs as above (C) and then passivating the discs for 30 min in 40 % nitric acid (ASTM F 86-01). Finally, the Heat Treated discs (HT) were obtained by taking discs that had been cleaned and passivated (C + P) and heat treating them at 350°C for 1 h in air, followed by air cooling.

Microstructure

Polished samples were etched using Kroll etchant (2%HF-4%HNO3) to reveal microstructures, and optical microscopy, and imaging camera equipment was used to examine and document the microstructures.

Auger Electron Spectroscopy (AES)

An AES system (JOEL, JAMP-30 Auger Microprobe) was used to evaluate the chemical composition of the prepared surfaces. Parameters used in the analysis included a 30° take-off angle, 10 keV accelerating electron beam, and 3.0×10^{-7} A probe current. Derivative spectra were used to analyze the elemental composition, with spectra being taken both before (pre-sputter) and after (post-sputter) Argon (Ar) ion sputtering. Continuous depth profiling was performed using Ar ions with a potential of 3 keV and emission current of 30 mA. The sputter rate of 0.2 nm/s was determined by sputtering through a standard 100 nm thick Ta_2O_5 film. The oxide thickness was determined by multiplying the sputter rate of 0.2 nm/s by the time it took the oxygen Auger signal to reach one-half its maximum intensity.

Potentiodynamic Polarization Corrosion Tests

Electrochemical potentiodynamic polarization testing was conducted using equipment (Model 273 Potentiostat/Galvanostat, EG&G PAR) and a corrosion cell standardized to ASTM G5. The polarization scan was conducted from 150 mV more active than open circuit potential to 1000 mV using a scan rate of 0.17 mV/s. A platinum counter electrode and standard calomel reference electrode (SCE) were used in this study. The electrolyte solution was isotonic saline solution adjusted to pH 7.4 ± 0.1. Tafel extrapolation and Stern-Geary fits (SoftCorr III, EG&G PAR) were used to obtain the corrosion potential (Ecorr) and the corrosion rate (Icorr) at the maximum corrosion potential. The passive current density (Ipass) was obtained by using a straight-line extrapolation to intersect the current density axis.

Electrochemical Impedance Spectroscopy (EIS)

Electrochemical impedance spectroscopic analyses were conducted using equipment (Model 6310 Impedance Analyzer, EG&G PAR) and a corrosion cell standardized to ASTM G 106. A platinum counter electrode and standard calomel reference electrode (SCE) were used in this study. A sinusoidal waveform with a frequency sweep from an initial value of 100 kHz to a final value of 10 MHz, with a magnitude of 10 mV, was used. The electrolyte solution was isotonic saline solution adjusted to a pH of 7.4 ± 0.1. A Randles equivalent circuit was fit to the impedance spectra (Model 398 Impedance Software EG&G PAR) and its circuit element values calculated. A linear relationship between the oxide film and the calculated capacitance is of the form:

$$C = \varepsilon o \ \varepsilon r \ A / d \qquad (1)$$

where

C = calculated double layer capacitance
A = nominal surface area of the specimen
d = thickness of the oxide film
εo = permittivity of free space
εr = specific permittivity of the oxide film (~100)

This formula was used to obtain relative differences in the oxide film thickness between the various sample groups. A certain magnitude of error is to be expected for this fit and analysis because the high corrosion resistance of titanium and its alloys in the test electrolyte precludes obtaining a Nyquist semicircle. Thus, the results obtained provide only qualitative information, not quantitative information.

Statistics

Duncan's Multiple Range Test (α = 0.05) was used to test for significant statistical differences, if any, in mean oxide thickness values and the corrosion parameters, Ecorr, Icorr, and Ipass.

Results

Microstructures

Representative photomicrographs show the microstructures of cpTi, Ti64, and 21SRx (Figs. 1–3). The cpTi grain structure was an equiaxed alpha phase, Fig. 1. The Cleaned, Passivated, and Heat Treated conditions showed similar microstructures. The Ti64, Fig. 2, was an alpha-beta microstructure with grains of alpha (light) and intergranular beta phase (mottled or outlined). No differences in microstructure morphology or grain size were found for the Cleaned, Passivated, and Heat Treated Ti64 groups. The microstructure of the 21SRx showed large equiaxed beta grains on a relative basis, Fig. 3. The larger beta phase for the 21SRx indicates that the bar stock from which the samples were made had been annealed. As with cpTi and Ti64, the Cleaned, Passivated, and Heat Treated 21SRx groups showed no differences in microstructure.

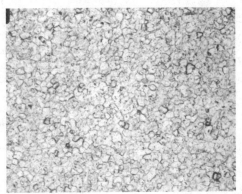

FIG. 1—*Representative microstructure for cpTi showing equiaxed alpha grains. Original magnification = 200×.*

FIG. 2—*Representative microstructure for Ti64 showing dual phase alpha-beta grains. Alpha regions are light and intergranular beta regions are dark. Original magnification = 200×.*

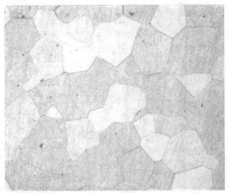

FIG. 3—*Representative microstructure for 21SRx showing equiaxed beta grains. Grains are substantially larger than cpTi and Ti64 grains. Original magnification = 200×.*

AES – Surface Oxide Elemental Composition

Pre-sputter AES spectra for all sample groups showed dominant Ti and O peaks and smaller C, Ca, P, and Si peaks. Minor Al and Mo peaks were detected throughout the surface oxides of the Ti64 and 21SRx specimens, respectively. Post-sputter spectra obtained after depth profiling showed dominant Ti peaks with much smaller O peaks and the disappearance of the C, Ca, and Si peaks that were observed in the pre-sputter spectra. In the case of the alloys, Ti64 and 21SRx, post-sputter spectra showed larger and more defined Al and Mo peaks, respectively. Figure 4 shows a representative post-sputter spectra for a Passivated 21SRx sample.

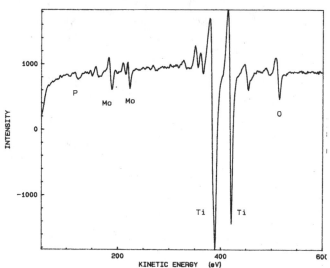

FIG. 4—*Representative AES spectra of Passivated 21SRx sample after Ar ion sputter.*

AES – Surface Oxide Thickness

The average oxide thicknesses as determined by AES continuous depth profiling are shown in Table 3.

TABLE 3—*Average oxide thickness (nm), n = 2.*

Surface Treatment	CpTi	Ti64	21SRx
Cleaned	3.0 ± 0.3 [a]	5.0 ± 0.3 [a]	4.3 ± 0.6 [a]
Passivated	4.4 ± 0.6 [a]	5.4 ± 0.3 [a]	5.0 ± 0.3 [a]
Heat Treated	16.8 ± 2.8 [c]	14.8 ± 0.3 [c]	8.8 ± 0.0 [b]

Results with different alphabetical superscripts denote statistically significant differences (Duncan's Multiple Range Test, α = 0.05).

All Cleaned and Passivated groups showed no significant differences (Duncan's Multiple Range Test, α = 0.05) in oxide thickness. The Heat Treated group oxides were significantly thicker than the Cleaned and Passivated groups, and within the Heat Treated groups, cpTi and Ti64 specimens had significantly thicker oxides than 21SRx specimens.

DC Corrosion Tests

The polarization corrosion curves were typical for titanium and titanium alloys, showing active and passive regions. Polarization curves for cpTi, Ti64, and 21SRx exhibited minimal differences in the locations or shapes of the curves within each specific surface treatment. However, differences were seen between surface treatment groups. Figures 5–7 show representative curves for the three surface treatments for each metal or alloy group. These figures illustrate graphically the effect of the surface treatments on each metal and alloy group. For all of the material groups, the Passivated and Heat Treated curves are shifted up and to the left relative to the Cleaned curves. This is indicative of increased corrosion resistance, with a shift up indicating an increase in Ecorr values (becoming more noble) and the shift to the left indicating smaller Icorr values (decreased corrosion current density). The curves for the Cleaned specimen groups showed passive regions in which the current densities were voltage-independent, whereas the Passivated and Heat Treated groups showed voltage-dependent passive regions.

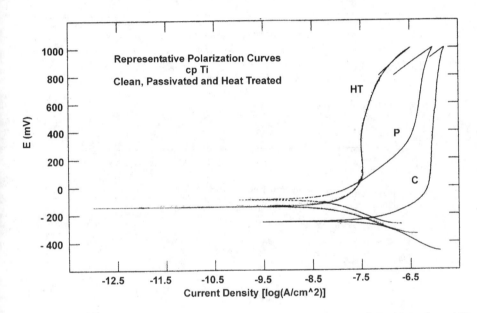

FIG. 5—*Representative polarization curves for cpTi in the Cleaned, Passivated, and Heat Treated conditions. The upward and leftward shift of the Passivated and Heat Treated curves relative to the Cleaned curve graphically illustrates their increased corrosion resistance. The upward shift indicates a nobler Ecorr and the leftward shift a decrease in current density (Icorr).*

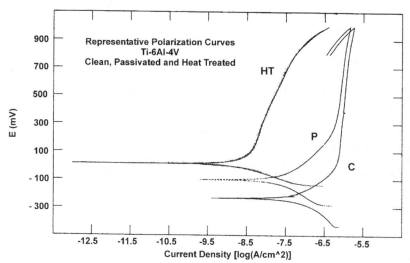

FIG. 6—*Representative polarization curves for Ti64 in the Cleaned, Passivated, and Heat Treated conditions. As with Fig. 5 for cpTi, the upward and leftward shift of the Passivated and Heat Treated curves relative to the Cleaned curve graphically illustrates their increased corrosion resistance. The upward shift indicates a nobler Ecorr and the leftward shift a decrease in current density (Icorr).*

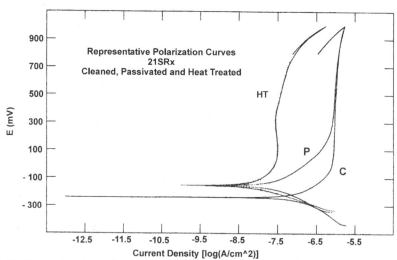

FIG. 7—*Representative polarization curves for 21SRx in the Cleaned, Passivated, and Heat Treated conditions. Although similar Cleaned and Passivated curves are shown, the quantitative data show the Passivated and Heat Treated groups to be significantly more corrosion resistant than the Cleaned group.*

Like the polarization curves, the Ecorr curve results were similar for all three material groups within the same treatment group but different between the treatment groups. The Cleaned and Heat Treated Ecorr curves showed gradual increases in potential over time, while the Passivated curves were relatively stable, as there was little to no change in the open circuit potential over the 1-h test period.

The average Ecorr, Icorr, and Ipass values for tested specimens are shown in Tables 4–6, respectively.

TABLE 4—*Average Ecorr results (mV), n = 3.*

Surface Treatment	cpTi Ecorr (mV)	Ti64 Ecorr (mV)	21SRx Ecorr (mV)
Cleaned	-259 ± 24 [a]	-235 ± 14 [a]	-250 ± 16 [a]
Passivated	-89 ± 14 [b]	-89 ± 40 [b]	-154 ± 49 [b]
Heat Treated	-123 ± 50 [b]	-9 ± 34 [c]	-115 ± 71 [b]

Results with different alphabetical superscripts denote statistically significant differences (Duncan's Multiple Range Test, $\alpha = 0.05$).

TABLE 5—*Average Icorr results (nA/cm^2), n = 3.*

Surface Treatment	cpTi Icorr (nA/cm^2)	Ti64 Icorr (nA/cm^2)	21SRx Icorr (nA/cm^2)
Cleaned	47 ± 2 [ab]	45 ± 14 [b]	61 ± 18 [a]
Passivated	6 ± 1 [c]	11 ± 3 [c]	13 ± 5 [c]
Heat Treated	5 ± 3 [c]	2 ± 1 [c]	12 ± 11 [c]

Results with different alphabetical superscripts denote statistically significant differences (Duncan's Multiple Range Test, $\alpha = 0.05$).

TABLE 6—*Average Ipass results (nA/cm^2), n = 3.*

Surface Treatment	cpTi Ipass (nA/cm^2)	Ti64 Ipass (nA/cm^2)	21SRx Ipass (nA/cm^2)
Cleaned	863 ± 119 [ab]	1101 ± 275 [a]	927 ± 63 [ab]
Passivated	541 ± 35 [c]	685 ± 94 [a]	859 ± 56 [ab]
Heat Treated	286 ± 251 [d]	9 ± 5 [c]	42 ± 9 [e]

Results with different alphabetical superscripts denote statistically significant differences (Duncan's Multiple Range Test, $\alpha = 0.05$).

The Ecorr and Icorr results from the polarization tests showed no significant differences between the metallic groups within any of the surface treatments, except for Cleaned Ti64, which had a significantly lower Icorr than Cleaned 21SRx alloy, and Heat Treated Ti64, which had a significantly lower Ecorr than Heat Treated cpTi and 21SRx. The surface treatment groups, however, showed significant differences in Ecorr and Icorr. The Cleaned specimens had significantly higher Icorr and less noble Ecorr values than the Passivated and Heat Treated groups. There was no significant difference in Ecorr and Icorr for the Passivated and Heat Treated groups, except for the previously mentioned Ecorr results for Heat Treated Ti64. The same general trends were shown for the Ipass results. The Cleaned specimen groups had the higher passive current densities, followed by the Passivated and then the Heat Treated groups. Significant differences were found only for the Heat Treated groups, with lower passive current densities than the Cleaned and Passivated groups.

Electrochemical Impedance Spectroscopy (EIS)

All impedance spectra fit into the Randles equivalent circuit model. All sample groups exhibited ideal capacitive behavior ($\phi \cong 90°$) expected for titanium and its alloys. The oxide thicknesses obtained for the various sample groups using EIS followed the same rank and file order as that obtained using AES.

Discussion

Microstructure

The microstructural analyses showed the three titanium materials to have three distinct grain structures, equiaxed alpha for all the cpTi groups, duplex alpha-beta for all the Ti64 groups, and equiaxed beta for all the 21SRx groups (Figs. 1–3). The analyses also showed that the substrate microstructure of each of the materials in the cleaned condition was not altered by the surface treatments used in this study, namely, nitric acid passivation and elevated temperature heat treatment (350°C for 1 h). The fact that the treatments did not alter the microstructures of the substrate materials was expected since nitric acid passivation is simply a chemical treatment of the surface, and the 350°C temperature used for the heat treatment is not sufficient to alter the grain structure of the titanium materials used in this study.

AES – Surface Oxide Chemistry

The AES results on the surface oxide chemistry from this study agree with results found in the literature for similar materials and surface treatments. For example, small peaks for carbon (C), calcium (Ca), phosphorus (P), and silicon (Si) were found for all the pre-sputter spectra, but disappeared after the Ar sputter, which suggests that these elements were part of a carboneous surface environment-dependent layer, which is routinely found on surfaces analyzed by AES [7]. The carboneous layer present on the surfaces was most likely due to adsorption of hydrocarbons from the cleaning solvents or air [7]. The other elements, Ca, P, and Si, were most likely the result of the grinding and polishing procedures [10]. As expected, both pre-sputter and post-sputter spectra showed large Ti and O peaks for all materials and surface treatments analyzed in this study. The Ti (LMV) peak at ~420 eV demonstrated a shape that corresponds closely to that of TiO$_2$ [26].

Spectra for the Ti64 alloy samples in the Cleaned, Passivated, and Heat Treated conditions were similar and showed peaks for Al but not V. Others have shown the presence of aluminum within the surface oxide. Vanadium, the other alloying element in Ti64, was not detected in any of the Ti64 alloy surfaces in this study. Other researchers have found vanadium within the surface oxide of Ti64 alloys for anodically prepared specimens [10] or when using x-ray photospectroscopy (XPS) [10,11,15]. One reason for not detecting vanadium in this study may be that the detection of vanadium in titanium alloys is difficult with AES since the vanadium AES peaks overlap those of titanium and oxygen [27].

The AES spectra for the Passivated 21SRx is shown in Fig. 4. Like the spectra for cpTi and Ti64, the 21SRx spectra show large, dominant titanium and oxygen peaks. All the 21SRx spectra also show small Mo peaks but no Nb. The presence of Mo in surface oxides of various Ti-Mo alloys has been reported using AES by other researchers [13,28]. Although no Nb peaks are shown in any of the Ti-15Mo-2.8Nb-0.2Si alloy samples, the presence of Nb cannot be completely ruled out since Nb peaks overlap Mo peaks.

Analysis of the AES spectra for nitrogen was not possible in this study because the only available nitrogen AES peak overlaps one of the Ti peaks [27]. Thus, any differences in nitrogen content of the material surfaces in this study could not be determined.

AES – Oxide Thickness

Oxide thicknesses for the different samples in this study were measured directly using AES depth profiling techniques. In addition, EIS analysis was used to measure the oxide thicknesses indirectly. AES and EIS analysis results were in basic agreement. Cleaned and Passivated groups for each of the materials in this study had similar oxide thicknesses ranging from 3.0–5.4 nm, with no significant differences between any of these groups. These results are in agreement with results from many other studies for similar materials and treatments [6,10,11,14]. These results, however, contradict Callen et al., who reported that the oxide thickness for nitric acid passivated Ti64 material was less than that of the non-passivated Ti64 [11]. Possible explanations include differences in sample preparation and analytical techniques used to measure oxide thickness. Another possibility is the concentration and nature of the nitric acid passivation used to treat the samples. Callen ultrasonicated the samples in a 34 % nitric acid solution, while in this study the samples were passivated statically in a 40 % nitric acid solution. During ultrasonication, high pressures and temperatures due to cavitations can occur near a metal surface. This, in turn, can affect chemical reactions and introduce a number of microstructural changes in the surface layers of the metal oxide [29]. Hence, nitric acid passivation done in conjunction with ultrasonication could alter the oxidation process during passivation and thus explain the thinner oxide results of the ultrasonically passivated samples found by Callen.

The heat treat regime used in this study was chosen so as to obtain relatively thin oxides, but ones that were significantly thicker than the native and passivated oxides. The 350°C-1 h regime was successful in doing this as Auger and EIS analysis showed that the 1-h, 350°C heat treatment significantly increased the thickness of the surface oxides of all three materials, compared to all the Cleaned and Passivated materials. Heat Treated cpTi and Ti64 had oxide thicknesses of 16.8 nm and 14.8 nm, respectively, which were not significantly different, while Heat Treated 21SRx had an average oxide thickness of 8.8 nm, which was significantly thinner than those of the Heat Treated cpTi and Ti64. The increased oxide thickness for the heat treatment regime used in this study agrees well with results reported in the literature [19,30,31]. In addition, all the Heat Treated materials were golden in color after the heat treatment, which has been reported in the literature to signify an oxide thickness of 10–25 nm [30], which agrees with the thickness results in this study.

The thinner oxide of the 21SRx Heat Treated group was attributed to the larger grain sizes of the alloy due to annealing (Fig. 3). Oxide growth in thermal treatments is controlled in large part by mass diffusion through the grain boundaries [32]. Thus, mass transport is greater in the fine-grained structures of the cpTi and Ti64 materials than in the larger-grained 21SRx, which leads to the thicker oxides for the fine-grained structured materials, cpTi and Ti64, compared to the larger-grained 21SRx.

DC Corrosion

In general, the corrosion results showed no significant differences between cpTi, T64, and 21SRx within specific treatment groups of this study. There were, however, significant differences between the treatment groups, with the Cleaned groups showing significantly less corrosion resistance than the Passivated and Heat Treated groups; the latter two, in general, did not show any significant differences.

However, the differences or lack of differences in corrosion resistance cannot be attributed simply to the thickness of the oxide layers. No significant differences in oxide thicknesses were found for the Cleaned and Passivated groups, yet the Passivated groups were significantly more corrosion resistant. On the other hand, the Heat Treated groups had significantly thicker oxides than the Passivated groups and yet, in general, showed no significant differences in corrosion resistance. These results suggest that other characteristics or properties of the oxide layers besides oxide thickness determine the corrosion properties of these materials.

One possible oxide property that may affect the corrosion properties is the presence and amount of sub-oxides in the oxides. Titanium sub-oxides such as TiO and Ti_2O_3 have been reported to be present in addition to stoichiometric TiO_2 on titanium surfaces [10,14,20,29,33]. Lausmaa [28] and Ong [14] reported the presence of sub-oxides in the oxide film for cpTi. Ask [10] reported sub-oxides of Ti_2O_3 and TiO for Ti64, while Milosev [33] reported oxide percentages of 40 % for TiO_2, 30 % for Ti_2O_3, and 4 % for TiO on Ti64. In addition, in a more recent study, Lee [34] quantitated the amount of titanium sub-oxides on Ti64 that had been surface treated in similar fashions as was done in this study. The XPS results showed the weight% of sub-oxides for "cleaned" Ti64 to be 19.5 wt%, while Ti64 that had been passivated in a 34 % nitric acid solution had 14.8 wt% of sub-oxides, and Ti64 that had been heat treated at 400°C for 1 h had 8.6 wt% of sub-oxides. Thus, these results show that nitric acid passivation and heat treatment decreased the amount of sub-oxides compared to the cleaned condition. This could explain the improved corrosion resistance for the Passivated and Heat Treated groups compared to the Cleaned groups in our study. However, more in-depth studies of the role(s) of oxide atomic structure are needed to draw conclusions about these relationships.

Another possible reason for the improved corrosion resistance of the Passivated groups compared to the Cleaned groups, despite the result that they had similar oxide thicknesses, might be the structure of the oxides. Nitric acid passivated specimens may remove the plastically deformed native oxide of the Cleaned specimens and grow in its place a more uniform oxide [35]. Since the nitric acid passivation process involves the simultaneous dissolution and formation of the existing oxide, preferential dissolution of surface defects would be expected, with concurrent formation of a less defective (i.e., more uniform) oxide.

Still another reason for the improvement in corrosion properties for the Passivated and Heat Treated groups may be enrichment of nitrogen in the near-surface region of the surface oxide due to the passivation in nitric acid. Schmidt [36] reported that ion-implanted nitrogen acts as a diffusion barrier by reducing the number of interstitial sites for migration of oxygen. Unfortunately, as discussed earlier, because the AES peak for titanium overlaps the nitrogen peak, changes in nitrogen content could not be determined.

Although the Heat Treated groups had increased oxide thicknesses, the Ecorr and Icorr results for these groups were not significantly different from those for the nitric acid Passivated groups. One possible explanation again may be the relative structure of the oxide. Blackwood [37] and others [38] have shown that the oxide layer on titanium and titanium alloys typically consists of three layers: an outer TiO_2 layer, an intermediate Ti_2O_3 layer, and an inner TiO layer that is in contact with the titanium or alloy substrate. In addition, Pouilleau [39] showed that a heat treatment similar to that done in this study resulted in a thin outer TiO_2 layer and a thicker intermediate Ti_2O_3 layer. Thus, the short and relatively low heat treatment temperature did not allow the formation of a thicker, more stable stoichiometric TiO_2 layer. The effect of the thinner TiO_2 layer on corrosion properties was shown after Pouilleau's oxide stabilization procedure of boiling in water for 15 min, which did not increase the overall thickness of the oxide layer but did increase the thickness of the outer TiO_2 layer, while decreasing the thickness of the intermediate Ti_2O_3 region. The stabilized specimens showed improved corrosion resistance, thus pointing to the importance of the different oxide structures. Hence, the result that the Heat Treated groups had a thicker oxide but not improved Ecorr and Icorr results compared to the Passivated groups may be due to the circumstance that the heat treatment used in this study was not of sufficient duration and temperature for a thick outer TiO_2 layer to form.

However, it should be noted that the Ipass results for the Heat Treated groups were significantly lower than those for the Passivated groups. Ipass, by definition, is the current density of the sample when it is in the passive region of the polarization curve. The lower Ipass results for the Heat Treated groups may be due to the possible transformation of the intermediate Ti_2O_3 sub-oxide layer into the more stable TiO_2 due to the application of the voltage during the polarization experiment. An increase in the amount of the more stable TiO_2 would result in a decrease in current density compared to the thinner Passivated groups.

Another result from the corrosion studies that provides information on the state of the oxides present on the samples is the shape of the Ecorr curves. Ecorr curves measure the open circuit potential over time. According to Abd El Kader [40], shifts in the open circuit potential to more positive values over time denote oxide thickening and repair, shifts to more negative values indicate oxide breakdown or dissolution, and a constant voltage over time indicates that the oxides were stable. In this study, all the Ecorr curves for the Cleaned and Heat Treated groups had voltages becoming slightly more positive with time, indicating that the oxide films were thickening and becoming repaired. On the other hand, all the Passivated groups had Ecorr curves exhibiting relatively constant voltages over time, indicating stable oxide films. Thus, nitric acid passivation appears to have stabilized the oxide in terms of its behavior in the saline solution, while Cleaned and Heat Treated materials had oxides that continued to passivate in the saline solutions, as indicated by the increasing open circuit potential.

Conclusions

The results of this study showed similar excellent corrosion properties for cpTi, Ti-6Al-4V (Ti64), and Ti-15Mo-2.8Nb-0.2Si (21SRx) materials despite differences in the chemistry of the surface oxides of these materials. Thus, surface oxide composition did not affect the corrosion properties of cpTi, Ti64, and 21SRx.

Surface treatments, such as passivation in nitric acid solution and heat treatment, did affect the corrosion properties by increasing the corrosion resistance of these surface-treated materials. However, these differences in corrosion properties cannot be attributed simply to changes in oxide thickness. Other oxide properties must play a role in the corrosion properties of these materials. Additional studies on the role(s) of atomic structure versus surface properties are indicated.

References

[1]   Fraker AC. "Corrosion of metallic implants and prosthetic devices," *Metals Handbook*. Metals Park, Ohio: ASM International; 1987.

[2]   Zitter H, Plenk H, Jr. "The electrochemical behavior of metallic implant materials as an indicator of their biocompatibility," *Journal of Biomedical Materials Research*, 1987, 21:881-896.

[3]   Williams DF. "Titanium and Titanium Alloys," In: Williams DF, ed. *Biocompatibility of*

Clinical Implant Materials. Boca Raton, Florida: CRC Press; 1981.

[4]   Turner TM, Sumner DR, Urban RM, Igloria R, Galante JO., "Maintenance of proximal cortical bone with use of a less stiff femoral component in hemiarthroplasty of the hip without cement: An investigation in a canine model at six months and two years," *Journal Bone Joint Surgery Am.,* 1997, 79:1381-1390.

[5]   Woo SL, Lothringer KS, Akeson WH, Coutts RD, Woo YK, Simon BR, Gomez MA. "Less rigid internal fixation plates: historical perspectives and new concepts," *Journal of Orthopecic Research,* 1984, 1:431-449.

[6]   Lausmaa J, Mattsson L, Rolander U and Kasemo B, "Chemical Composition and Mophology of Titanium Surface Oxides," *Materials Research Society Symposia Proceeding,.* 1986, 55:351.

[7]   Lausmaa J, Kasemo B, Mattsson H and Odelius H, "Multi-technique surface characterization of oxide films on electropolished and anodically oxidized titanium," *Applied Surface Science,* 1990, 45:189.

[8]   Hiromoto S, Hanawa T and Asami K, "Composition of surface oxide film of titanium with culturing murine fibroblasts L929," *Biomaterials,* 2004, 25:979.

[9]   Lausmaa J and Kasemo B, "Surface Oxides on Titanium Implants-Spectroscopic Studies of their Composition and Thickness, and Implications for the Biocompatibility of Titanium," *Transactions of Society for Biomaterials,* Society for Biomaterials, Minneapolis, MN, Vol. 7, 1984, p. 231.

[10]   Ask M, Lausmaa J and Kasemo B, "Preparation and surface spectroscopic characterization of oxide films on Ti6Al4V," *Applied Surface Science,* 1989, 35:283.

[11]   Callen BW, Lowenberg BF, Lugowski S, Sodhi RNS and Davies JE, "Nitric acid passivation of Ti6Al4V reduces thickness of surface oxide layer and increases trace element release," *Journal of Biomedical Materials Research,* 1995, 29:279.

[12]   Lausmaa J, Ask M, Rolander U, Kasemo B, Bjursten LM, Ericson LE and Thomsen P, "Surface preparation and spectroscopic analysis of titanium implant materials," *Surface and Interface Analysis,* 1990, 16:571.

[13]   Laser D and Marcus HL, "Auger Electron Spectroscopy Depth Profile of Thin Oxide on a Ti-Mo Alloy," *Journal Electrochemical Society,* 1980, 127:763-765.

[14]   Ong JL, Lucas LC, Raikar GN and Gregory JC, "Electrochemical corrosion analyses and characterization of surface-modified titanium," *Applied Surface Science,* 1993,72:7.

[15]   Sodhi RN, Weninger A and Davies JE, "X-ray photoelectron spectroscopic comparison of sputtered Ti, Ti6Al4V, and passivated bulk metals for use in cell culture techniques," *Journal of Vacuum Science Technolog,* 1991, A9:1329-1333.

[16]   Smith DC, Pilliar RM and Chernecky R, "Dental implant materials. I. Some effects of preparative procedures on surface topography," *Journal of Biomedical Materials Research,* 1991;25:1045.

[17]   Smith DC, Pilliar RM, Metson JB and McIntyre NS, "Dental implant materials. II. Preparative procedures and surface spectroscopic studies," *Journal of Biomedical Materials Research,* 1991, 25:1069.

[18]   Lowenberg BF, Lugowski S, Chipman M and Davies JE, "ASTM-F86 passivation increases trace element release from Ti6Al4V into culture medium," *Journal of Materials Science: Materials in Medicine,*1994, 5:467.

[19]   Kilpadi DV, Raikar GN, Liu J, Lemons JE, Vohra Y and Gregory JC, "Effect of surface treatment on unalloyed titanium implants: Spectroscopic analyses," *Journal of Biomedical Materials Research,* 1998, 40:128.

[20]   Lee TM, Chang E and Yang CY, "Surface characteristics of Ti6Al4V alloy: Effect of materials, passivation and autoclaving," *Journal of Materials Science: Materials in Medicine,* 1998, 9:439.

[21]   Browne M, Gregson PJ and West RH, "Characterization of titanium alloy implant surfaces with improved dissolution resistance," *Journal of Materials Science: Materials in Medicine,* 1996, 7:323.

[22]   Garcia-Alonso MC, Saldana L, Valles G, Gonzalez-Carrasco JL, Gonzalez-Cabrero J,

Martinez ME, Gil-Garay E and Munuera L, "In vitro corrosion behaviour and osteoblast response of thermally oxidised Ti6Al4V alloy," *Biomaterials,* 2003, 24:19.

[23] Kilpadi DV and Lemons JE, "Effect of surface and heat treatments on corrosion of unalloyed titanium implants," *Southern Biomedical Engineering Conference-Procedings,* IEEE, Piscataway, NJ, 1997, p. 70.

[24] Lee TM, Chang E and Yang CY, "Effect of passivation on the dissolution behavior of Ti6Al4V and vacuum-brazed Ti6Al4V in Hank's ethylene diamine tetra-acetic acid solution Part I ion release," *Journal of Materials Science: Materials in Medicine,* 1999, 10:541.

[25] Park YJ, Shin MS, Yang HS, Ong JL and Rawls HR, "Effect of heat treatment on microstructural and electrochemical properties of titanium and Ti alloy," *Southern Biomedical Engineering Conference-Procedings,* IEEE, Piscataway, NJ, 1998, p. 46.

[26] Lausma J, Ask M, Rolander U and Kasemo B, "Preparation and Analysis of Ti and Alloyed Ti Surfaces Used In The Evaluation of Biological Response," *Materials Research Society Symposia Proceedings,* 1989, 110:647-653.

[27] Lausmaa J, Kasemo B and Mattsson H, "Surface spectroscopic characterization of titanium implant materials," *Applied Surface Science,* 1990, 44:133.

[28] Kim YJ and Oriani RA, "Effect of the Microstructure of Ti-5Mo on the Anocic Dissolution in H2SO4," *Corrosion,* 1987, 43:418.

[29] Whillock GOH and Harvey BF, "Preliminary investigation of the ultrasonically enhanced corrosion of stainless steel in the nitric acid/chloride system," *Ultrasonics Sonochemistry,* 1996, 3:111-118.

[30] Velten D, Biehl V, Aubertin F, Valeske B, Possart W and Breme J, "Preparation of TiO(2) layers on cp-Ti and Ti6Al4V by thermal and anodic oxidation and by sol-gel coating techniques and their characterization," *Journal of Biomedical Materials Research,* 2002, 59:18-28.

[31] Chang E and Lee TM, "Effect of surface chemistries and characteristics of Ti6Al4V on the Ca and P adsorption and ion dissolution in Hank's ethylene diamine tetra-acetic acid solution," *Biomaterials,* 2002, 23:2917.

[32] Leyens C, Peters M and Kaysser WA, "Influence of Microstructure on Oxidation Behaviour of Near-Alpha Titanium Alloys," *Materials Science and Technology,* 1996, 12:213-218.

[33] Milosev I, Metikos-Hukovic M and Strehblow HH, "Passive film on orthopaedic TiAlV alloy formed in physiological solution investigated by X-ray photoelectron spectroscopy," *Biomaterials,* 2000, 21:2103.

[34] Lee TM, Chang E and Yang CY, "Comparison of the corrosion behaviour and surface characteristics of vacuum-brazed and heat-treated Ti6Al4V alloy," *Journal of Materials Science: Materials in Medicine,* 1998, 9:429.

[35] Trepanier C, Tabrizian M, Yahia LH, Bilodeau L and Piron DL, "Effect of modification of oxide layer on NiTi stent corrosion resistance," *Journal of Biomedical Materials Research,* 1998, 43:433.

[36] Schmidt H, Stechemesser G, Witte J and Soltani-Farshi M, "Depth distributions and anodic polarization behaviour of ion implanted Ti6Al4V," 1998, 40:1533.

[37] Blackwood DJ, Greef R and Peter LM, "Ellipsometric study of the growth and open-circuit dissolution of the anodic oxide film on titanium," *Electrochimica Acta,* 1989, 34:875.

[38] Pan J, Thierry D and Leygraf C, "Electrochemical impedance spectroscopy study of the passive oxide film on titanium for implant application," *Electrochimica Acta,* 1996, 41:1143.

[39] Pouilleau J, Devilliers D, Garrido F, Durand-Vidal S and Mahe E, "Structure and composition of passive titanium oxide films," *Materials Science & Engineering B: Solid-State Materials for Advanced Technology,.* 1997, B47:235.

[40] Abd El Kader JM, Abd El Wahab FM, El Shayeb HA and Khedr MGA, "Oxide Film Thickening on Titanium in Aqueous Solutions in Relation to Anion Type and Concentrtion," *British Corrosion Journal,* 1981, 16:111.

Journal of ASTM International, July/August 2005, Vol. 2, No. 7
Paper ID JAI12809
Available online at www.astm.org

Lyle D. Zardiackas,[1] *Michael D. Roach,*[1] *and R. Scott Williamson*[1]

Comparison of Stress Corrosion Cracking and Corrosion Fatigue (Anodized and Non-Anodized Grade 4 CP Ti)

ABSTRACT: In light of the possible effects of anodization on stress corrosion cracking (SCC) and corrosion fatigue (CF) of CP titanium, a research project has been completed recently in our laboratories to evaluate and compare SCC and CF of anodized versus non-anodized samples from a single lot of Grade 4 CPTi. Initial evaluation of alloy composition, microstructure, Vickers microhardness, and mechanical properties including the tensile and yield strength, % elongation and reduction of area was performed. After these tests ensured the material met the standards of ASTM Standard Specification for Unalloyed Titanium for Surgical Implant Applications (F 67), samples were prepared using low stress grinding techniques. Samples were divided into two groups, and the non-anodized SCC and CF testing was initiated. The surface of the second group of samples was anodized by Synthes to provide a green/gold surface consistent with standard production processing and then tested using the identical methodology as used for the non-anodized samples.

Results of the slow strain rate SCC testing on both smooth and notched anodized and non-anodized samples in both distilled de-ionized water and Ringers solution at 37°C showed no effect of anodization. Results of corrosion fatigue testing indicated that, while there was no effect of anodization on corrosion fatigue, there was a significant effect of the notch on the fatigue characteristics regardless of the two different surface conditions.

KEYWORDS: anodization, stress corrosion cracking, corrosion fatigue, notch sensitivity, CP titanium

Introduction

Over the years, there has been a great deal of conjecture and controversy surrounding the stress corrosion cracking and corrosion fatigue behavior of alloys used for implants. Much of the controversy has involved the potential for the occurrence of both of these failure modes in 316L stainless steel and in certain titanium alloys in the presence of physiological saline and at body temperatures. There were a number of papers published in the 1960s and 1970s discussing these phenomena in retrieved implants [1–7] as well as laboratory research in this area [8–10]. However, the vast majority of all of these studies examined 316 and 316L stainless steel implants, and none reported significant findings related to CP titanium or titanium alloys. Since the changes in composition of implant quality 316L as required by ASTM Standard Specification for Wrought 18 Chromium-14 Nickel-2.5 Molybdenum Stainless Steel Bar and Wire for Surgical Implants (F 138), the observations have substantially changed and in general do not include observation of SCC on retrieved 316L implants.

Early research by Scully [11], presented at the International Symposium on Stress Corrosion Mechanisms in Titanium Alloys in 1971, suggested that the appearance of cleavage type fractures in CP titanium is due to SCC and that fluting and striations in certain cases could be

Manuscript accepted for publication 7 January 2005; published July 2005. Presented at ASTM Symposium on Titanium, Niobium, Zirconium, and Tantalum for Medical and Surgical Applications on 9-10 November 2004 in Washington, DC.
[1] Professor and Chair Department of Biomedical Materials Science, Senior Materials Engineer, Senior Materials Engineer, University of Mississippi Medical Center, Jackson, MS 39216.

synonymous. However, since that time it has been well established that fluting and striations are due to entirely different phenomena [12–15]. Research by Bundy [16] suggested SCC as a possible contributory mechanism for the fracture of CP titanium and Ti-6Al-4V ELI. Since the crack propagation phase of CF (Stage II) and SCC (dynamic) are often considered to be the same mechanism, and since the fatigue fracture surface morphology of CP titanium is highly dependent on stress intensity levels, with striations, fluting, smooth terraces, and terraces with feather marks possible, it is often difficult, if not impossible, to determine if the fracture mechanism in a specific area of the fracture surface is fatigue or stress corrosion cracking [17–19]. Additionally, this controversy is a moot point once the crack has formed, since propagation of the crack will ensue provided that the stresses remain and the implant's function will continue to diminish to final fracture. These researchers' studies, as well as others, have defined fatigue as the method of fracture of titanium.

Most of the fundamental research in the area of SCC and CF of CP titanium and titanium alloys has been conducted using solutions of 3.5 % Sodium Chloride (NaCl) or other solutions with high Cl⁻concentrations to simulate sea water. This research has shown, and there is general agreement, that CP titanium with greater than 0.3 % oxygen is susceptible to SCC in seawater under certain conditions. Much of this research has focused on the effects of oxygen concentration on the crack propagation phase after the formation of a fatigue crack and is not related to crack initiation [22–24]. There has, however, been disagreement concerning the susceptibility of CP titanium to SCC in vivo where the concentration of Cl⁻ is significantly lower.

Over the years, questions have been raised concerning the effect of anodization on the SCC and CF of CP titanium since anodization in other alloy systems, especially certain aluminum alloys, can affect these properties negatively [25]. It is well recognized that anodization of titanium can have significant negative effects on properties, unless the composition of the acid used for the acid activation step is carefully controlled [26,27]. If the ratio of nitric to hydrofluoric acid does not remain greater than 10/1, there may be sufficient hydrogen uptake to cause embrittlement.

There are two generally accepted methodologies that may be employed when evaluating SCC [28–31] of samples that have not been subjected to a fatigue pre-crack. The first method uses a constant strain rate. To accurately perform this test, an extensiometer or strain gauge must be affixed to the sample and remain in solution for the duration of the test. This regimen presents a number of significant problems due to possible problems with gauge retention on the sample, the effects of corrosion on the gauges, the possible galvanic couples between the gauges, etc. The second method uses a constant extension rate [ASTM Standard Practice for Slow Strain Rate Testing to Evaluate the Susceptibility of Metallic Materials to Environmentally Assisted Cracking (G 129)]. In this method, only the sample is in the corrosive environment. While the strain rate is not constant, if the strain rate is sufficiently low, the small change in strain rate over the duration of the test will not adversely affect the results. Since both methodologies are acceptable, and there are difficulties involved without the use of an extensiometer, the constant extension rate method was used

The fatigue behavior of materials may be evaluated using different methodologies dependent on the information sought. These methods may be grouped into those used to determine statistically valid fatigue load or stress limits at a specific number of cycles and those designed to generate a S/N curve or Woehler diagram. Since the goal of this research was to compare and contrast the fatigue behavior of anodized versus non-anodized CP titanium over a range of load levels, the method of generating S/N curves for each condition was chosen. There are also several acceptable methods to develop S/N fatigue curves for metals. Among these are rotating beam, cantilever, three and four point bending, tension-compression, and tension-tension. Since there is an accepted method within ASTM F 04 (ASTM F 1801), and since we have previously characterized the fatigue behavior of several other implant alloys using this methodology, this method was followed.

The goal of this research was to compare the SCC behavior of smooth and notched Grade 4 CP Ti with and without anodization, and to evaluate the CF behavior of smooth and notched samples of Grade 4 CP Ti with and without anodization. It was hypothesized that there is no difference in the SCC or CF behavior as a function of anodization regardless of whether smooth or notched samples were evaluated.

Materials and Methods

Annealed Grade 4 CP titanium with an oxygen content of 0.38 % was supplied as 8 mm round from a single lot. Samples were machined at Metcut using low stress grinding techniques. Areas within the gauge or notch had a surface finish of Ra = 16 μin or better (Fig. 1). For notched samples, dimensions were determined to achieve a K_t factor of 3.2. This value was chosen since much of the work on notched samples cited in the literature and all of the research performed in our laboratories on SCC and CF of notched samples has been on samples with a K_t value of 3.2. After machining and inspection, one-half of the samples from each group was anodized by Synthes[2] and re-inspected. Since the thickness of the anodized layer was in the neighborhood of 130 nm, the K_t factor (3.2) of the notched samples after anodization was not significantly affected.

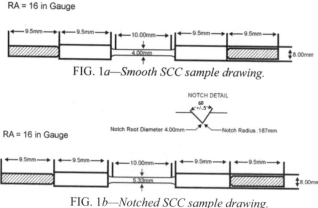

FIG. 1a—Smooth SCC sample drawing.

FIG. 1b—Notched SCC sample drawing.

For reasons previously discussed, SCC testing was performed according to the slow strain rate method (ASTM G 129). Smooth and notched tensile samples with and without anodization were tested in both distilled H_2O and Ringers solution at 37°C. Testing was performed [1] using an MTS servo-hydraulic testing system at 10^{-5} mm·sec^{-1}, and samples were maintained in solution at 37°C to fracture. Data, including load and deformation, were stored in the system's computer.

---

[2] Synthes (USA), 1301 Goshen Parkway, West Chester, PA 19380.

Smooth and notched samples for CF were tested since there has been other research, primarily using the rotating beam methodology, showing a notch affect in fatigue. Samples were prepared in an identical manner to that described under the section on SCC, except that the diameter in the gauge and at the root of the notch was 2.5 mm, and the notch dimensions were altered to maintain a K_t factor of 3.2. Samples were cycled in tension-tension in aerated Ringer's solution in a controlled temperature chamber at 37°C using an MTS servo-hydraulic testing system, following the guidelines of ASTM Standard Practice for Corrosion Fatigue Testing of Metallic Implant Materials (F 1801). The load was applied as a sinusoidal wave function in load control at 1 hz. A minimum of three samples was tested at each of five load levels including run-out at 10^6 cycles to generate S/N curves for each of the conditions. Analysis of fracture surfaces was performed using scanning electron microscopy (SEM).

Results

Stress Corrosion Cracking

Results of testing smooth and notched SCC samples with the two surface conditions (anodized and non-anodized) in Ringer's and distilled/de-ionized (DI) water, at 37°C are given in Table 1. Percentage elongation and reduction of area values on the smooth samples revealed only minor, if any, differences for both anodized and non-anodized samples between the Ringer's solution and DI water. To determine the effects of anodization on ductility of the anodization process, percent elongation ratios (PER) and reduction of area ratios (ROAR) were calculated. These ratios are calculated as the mean value of percent elongation or reduction of area of the sample set representing the test group, divided by the mean value of percent elongation or reduction of area of the sample set representing the control group. As an example, the mean percentage elongation of anodized samples divided by the mean percentage elongation of non-anodized samples tested in the same solution.

Representative micrographs of SEM analysis of the smooth SCC samples are seen in Figs. 2a–c. Analysis of the smooth samples revealed ductile cup and cone fractures (a and b) with dimples evident at higher magnifications (c). No significant differences were found in what would be considered typical ductile fracture morphology, even for samples tested in Ringer's solution.

TABLE 1—*SCC Results – CP Titanium – Grade 4.*

Sample Condition	Solution	% EL	ROA
Smooth	Ringer's	14.76 ± 0.03	40.53 ± 1.96
Smooth	DI H$_2$O	15.08 ± 0.47	41.22 ± 1.40
Smooth (a)	Ringer's	15.36 ± 0.19	40.61 ± 2.10
Smooth (a)	DI H$_2$O	15.31 ± 0.11	40.55 ± 0.39
Notched	Ringer's	19.33 ± 0.42	9.33 ± 0.54
Notched	DI H$_2$O	20.00 ± 1.75	10.23 ± 0.54
Notched (a)	Ringer's	20.93 ± 11.05	11.05 ± 0.79
Notched (a)	DI H$_2$O	20.06 ± 0.70	10.70 ± 0.25

a = anodized samples.

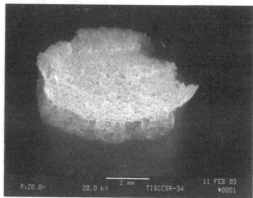

FIG. 2a—*Fracture surface of smooth CP Ti SCC sample in Ringer's.*

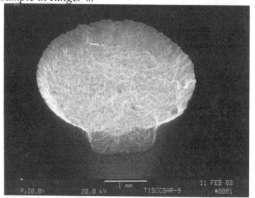

FIG. 2b—*Fracture surface on anodized smooth CP Ti SCC sample in Ringer's.*

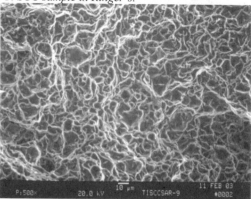

FIG. 2c—*Ductile overload dimples on anodized smooth CP Ti SCC sample in Ringer's.*

For the notched samples, calculation of PERs or ROARs for the notched non-anodized condition showed essentially no effect. These results were also observed for the ratios of the notched anodized group when tested against the non-anodized notched group in distilled de-ionized water (Table 2). However, for the notched anodized group/non-anodized samples in Ringer's, a PER of 109.13 and ROAR of 118.40 were calculated. This ratio indicates slightly less susceptibility to SCC for the notched anodized samples as compared to the notched non-anodized samples when tested in Ringer's solution.

SEM analysis revealed a similar ductile cup and cone fracture with dimples to that of the smooth samples (Figs. 3a–c). However, due to the change in stress concentration at the root of the notch, prominent secondary cracking of the fracture surfaces was also evident on all samples (Figs. 4a and 4b).

TABLE 2—*SCC Results – CP Titanium – Grade 4 PER and ROAR.*

Sample Condition	Solution Comparison	PER	ROAR
Smooth	Ringer's / DI H_2O	97.90	98.32
Smooth	Ringer's anodized /Ringer's	104.02	100.19
Smooth	DI H_2O anodized / DI H_2O	101.49	98.37
Notched	Ringer's / DI H_2O	96.65	91.22
Notched	Ringer's anodized / Ringer's	109.13	118.40
Notched	DI H_2O anodized / DI H_2O	100.79	104.56

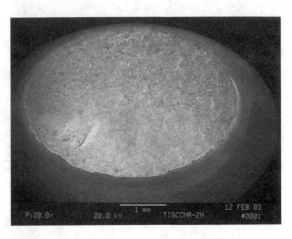

FIG. 3a—*Fracture surface on notched CP Ti sample in Ringer's—note secondary cracking.*

FIG. 3b—*Fracture surface on anodized notched CP Ti SCC sample in Ringer's.*

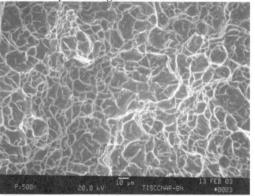

FIG. 3c—*Ductile dimpled fracture on anodized notched CP Ti SCC sample in Ringer's.*

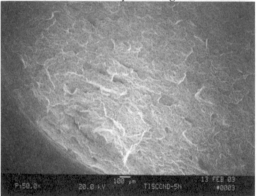

FIG. 4a—*Secondary cracking on fracture surface of notched SCC sample in DI water.*

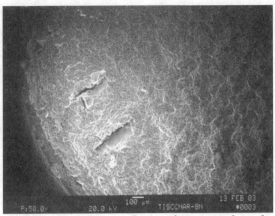

FIG. 4b—*Secondary cracking on fracture surface of anodized notched SCC sample in Ringer's.*

Corrosion Fatigue

Stress versus number of cycles to failure (S/N) curves were generated for all of the tested conditions as shown in Figs. 5a (smooth samples) and 5b (notched samples). Materials are considered to be susceptible to corrosion fatigue if the S/N curve in an aggressive medium is significantly depressed compared to the S/N curve in a non-aggressive medium. No significant differences were apparent in the S/N curves generated in Ringer's and DI water for the smooth anodized and non-anodized samples or between the notched anodized and non-anodized samples. These results indicate that there is little, if any, susceptibility to corrosion fatigue for CP Ti Grade 4 under either of the surface conditions (anodized versus non-anodized). SEM analysis confirmed this with similar fracture mechanisms for the two surface conditions (Figs. 6 and 7). Each of the samples showed fracture surfaces with clearly defined areas of fatigue and a shear lip in the area of final fracture (Figs. 6a and 7a). Crack initiation was clearly defined by chevron marks as being initiated at the surface of each sample (Figs. 6b and 7b). Most of each of the fracture surfaces exhibited typical Stage II fatigue crack propagation with striations and some secondary cracking (Figs. 6c and 7c) as well as terraces with feather marks and fluting (Figs. 6c and 7d). The presence of the terraces and fluting was more pronounced on samples tested at higher stress levels. Final fracture in the areas of the shear lips was corroborated by the presence of dimples in these areas (Figure 6d).

Results of this portion of the study showed no significant differences in the S/N curves for the notched samples, regardless of surface condition. As expected from other work in our laboratories, as well as the work of others [32,33], the fatigue curves for notched samples were significantly shifted to lower stress levels at corresponding numbers of cycles to failure when compared to those of smooth samples tested under the same conditions (Figs. 5a and 5b). Once again, SEM analysis confirmed the data, as the fracture morphologies were essentially the same regardless of surface condition (anodized versus non-anodized) when tested in the two solutions. The morphology over most of the fracture surface of samples tested at lower peak stress was similar to that of the smooth samples with finely spaced striations (Fig. 8), some secondary cracking, terraces, fluting, and smeared metal in areas where opposing fracture surface contacted during and after crack propagation. As the peak load was reduced, resulting in a greater number of cycles to failure, a significantly different fracture pattern was observed. The fracture surfaces exhibited a significantly greater amount of fluting and terracing along some widely spaced striations consistent with low cycle Stage II fatigue crack propagation (Figs. 9 and 10). Final ductile overload fracture occurred toward the middle of the fracture surface.

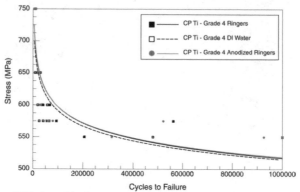

FIG. 5*a*—*S/N fatigue curves for smooth CP Ti Grade 4.*

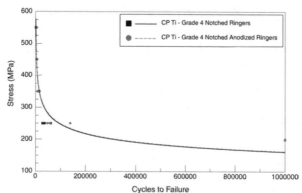

FIG. 5*b*—*S/N fatigue curves for notched CP Ti Grade 4.*

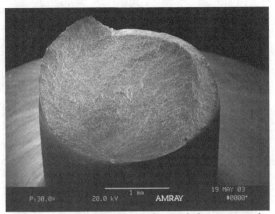

FIG. 6a—*Fracture surface of smooth fatigue sample in Ringer's at low stress (550 MPa, 205 340 cycles).*

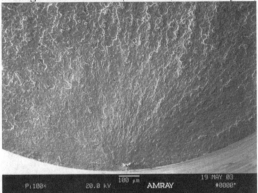

FIG. 6b—*Crack initiation region of Sample in Fig.* 6a.

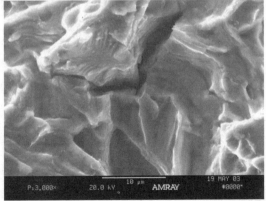

FIG. 6c—*Striations, flutes, terraces with feather marks, and secondary cracking on sample in* a.

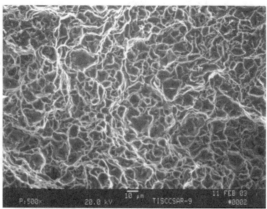

FIG. 6d—*Dimples in area of shear lip in Fig. 6a.*

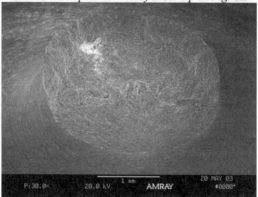

FIG. 7a—*Smooth anodized fatigue fracture surface in Ringer's at low stress (550 MPa, 926 560 cycles).*

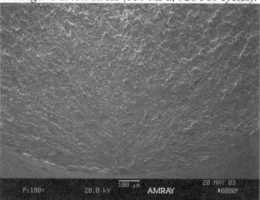

FIG. 7b—*Crack initiation region of fracture surface in Fig. 7a.*

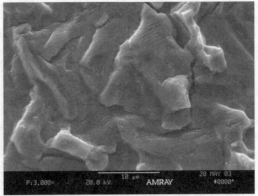

FIG. 7c—*Fatigue striations and secondary cracking on fracture surface in Fig. 7a.*

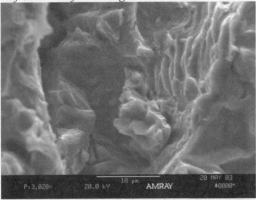

FIG. 7d—*Fluting and terraces with feather marks and secondary cracking on fracture surface in Fig. 7a.*

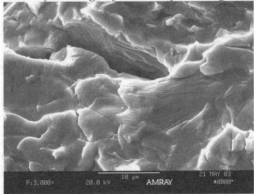

FIG. 8—*Striations and secondary cracks on anodized notched sample in Ringer's at low stress (138 474 cycles).*

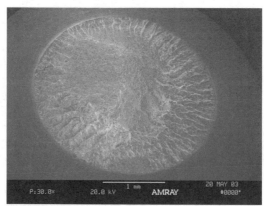

FIG. 9a—*Fracture surface of notched fatigue sample in Ringer's at high stress (550 MPa, 1023 cycles).*

FIG. 9b—*Fluting and terraces with feather marks prevalent on fracture surface in 9a.*

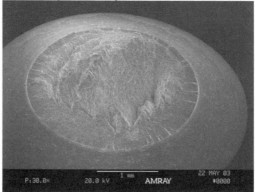

FIG. 10a—*Fracture surface of anodized notched fatigue sample in Ringer's at high stress (1840 cycles).*

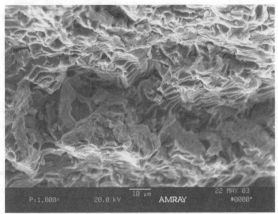

FIG. 10*b*—*Extensive fluting and terracing on fracture surface of sample in 10*a.

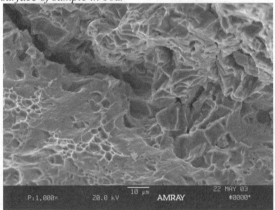

FIG. 10*c*—*Transition from Stage II fatigue to ductile overload on sample in 10*a.

Discussion and Conclusions

The results of this study have shown that under the anodizing conditions used, there is no increase in the susceptibility to SCC and CF of anodized annealed Grade 4 CP titanium, as compared to non-anodized annealed Grade 4 CP titanium. On the contrary, under the conditions evaluated, surface anodization reduced the susceptibility to SCC. It is highly possible that the presence of this very thin anodized surface ($\approx$0.130 μm) reduces the susceptibility for crack initiation, whereas in aluminum alloys the anodized layer is on an order of magnitude greater in thickness (>1.0μm). Because anodization creates an oxide surface layer that is brittle and is much thicker for aluminum, it is likely that there is a tendency for it to crack easily. It is also interesting to note that no fracture morphology specific to SCC was noted and the fracture surface morphology on samples tested in DI water and Ringers were essentially identical even though the oxygen content exceeded 0.30 %. While the results of research for this material tested at higher concentrations of Cl⁻, these results indicate that under the conditions tested in this research, Grade 4 CP titanium with up to 0.38 % oxygen is not susceptible to SCC.

The effect of anodization has likewise been shown to have no adverse effect on the fatigue behavior of Grade 4 CP titanium under the conditions tested. However, the presence of a notch, regardless of whether samples were anodized or non-anodized, has been shown to adversely affect the fatigue characteristics of this alloy. These observations on the notch sensitivity in fatigue are in direct agreement with others who have examined evaluated CP titanium as well as other titanium alloys used for biomedical applications.

There is another methodology for evaluation of SCC, often called environmentally assisted cracking (EAC), which involves the use of fatigue pre-cracked fracture toughness type specimens [ASTM Standard Test Method for Determining a Threshold Stress Intensity Factor for Environment-Assisted Cracking of Metallic Materials (E 1681)]. This methodology is used to determine the crack propagation rate and assumes that the existence of a fatigue pre-crack exacerbates the problem of SCC because of the stress concentration and electrochemistry at the crack tip. While this may or may not be the case for titanium and its alloys under physiological conditions, implanted devices usually fracture under conditions of high cycle fatigue where the formation of a fatigue crack may take up to 90 % of the total cycles to failure. Therefore, whether or not there is an increase in the crack propagation rate is probably insignificant if not irrelevant. However, in light of these findings, further research regarding the effects of fatigue pre-cracking on SCC (ASTM E 1681) of CP titanium, alpha-beta alloys such as Ti-6Al-4V and Ti-6Al-7Nb, and a beta alloy (Ti-15Mo) is underway in our laboratories. Additionally, due to the apparent affect of a notch on the fatigue of CP titanium, further research concerning notch affects on the aforementioned alloys appears warranted and advisable.

Acknowledgments

This research was funded by a grant from Synthes.

References

[1] Cahoon, J. R. and Paxton, H. W., "Metallurgical Analyses of Failed Orthopedic Implants," *Journal of Biomedical Materials Research*, Vol. 2, 1968, pp. 1–22.
[2] "Bioceramics – Engineering in Medicine (Part 2)," C.W. Hall, S.F. Hulbert, S.N. Levine, and F.A. Young, Editors, *Biomedical Materials Symposium No. 2, Engineering Foundation Research Conferences*, New England College, Henniker, NH, August 3–7, 1970.
[3] Colangelo, V. J., "Corrosion Fatigue in Surgical Implants," *Journal of Basic Engineering*, December 1969, pp. 581–586.
[4] Hughes, A. N. and Jordan, B. A., "Metallurgical Observations on Some Metallic Surgical Implants Which Failed in Vivo," *Journal of Biomedical Materials Research*, Vol. 6, 1972, pp. 33–48.
[5] Dobbs, H. S. and Scales, J. T., "Fracture and Corrosion in Stainless Steel Total Hip Replacement Stems," *Corrosion and Degradation of Implant Materials Symposium, ASTM STP 684*, B. C. Syrett and A. Acharya, Es., ASTM International, West Conshohocken, PA, 1979, pp. 245–258.
[6] Brunner, H. and Simpson, J. P., "Fatigue Fracture of Bone Plates,"*Injury: The British Journal of Accident Surgery*, Vol. 11, No. 3, pp. 203–207.

[7] Gray, R. J., "Metallographic Examination of Retrieved Intramedullary Bone Pins and Bone Screws from the Human Body," Contract No. W–7405–eng–26, Oak Ridge National Laboratory, February 1973.

[8] Bundy, K. J. and Desai, V. H., "Studies of Stress-Corrosion Cracking Behavior of Surgical Implant Materials Using a Fracture Mechanics Approach," *Corrosion and Degradation of Implant Materials: Second Symposium, ASTM STP 859*, A. C. Fraker and C .D. Griffin, Eds., ASTM International, West Conshohocken, PA, 1985, pp. 73–90.

[9] Imam, M. A., Fraker, A. C., and Gilmore, C. M., "Corrosion Fatigue of 316L Stainless Steel, Co–Cr–Mo Alloy, and ELI Ti–6Al–4V," *Corrosion and Degradation of Implant Materials Symposium, ASTM STP 684*, B.C. Syrett and A. Acharya, Eds., ASTM International, West Conshohocken, PA, 1979, pp. 128–143.

[10] Sheehan, J. P., Morin, C. R., and Packer, K. F., "Study of Stress Corrosion Cracking Susceptibility of Type 316L Stainless Steel in Vitro," *Corrosion and Degradation of Implant Materials: Second Symposium, ASTM STP 859*, A. C. Fraker and C. D. Griffin, Eds., ASTM International, West Conshohocken, PA, 1985, pp. 57–72.

[11] Scully, J. C., "Failure Analysis of Stress Corrosion Cracking with the Scanning Electron Microscope," *An International Symposium on Stress Corrosion Mechanisms in Titanium Alloys*, Georgia Institute of Technology, Atlanta, GA, January 27–29, 1971.

[12] Chesnutt, J. C. and Williams, J. C., "Comments on the Electron Fractography of α-Titanium," *Metallurgical Transactions A*, Vol. 8A, March 1977, pp. 514–515.

[13] Judy, R. W., Caplan, I. L., and Bogar, F. D., "Effects of Oxygen and Iron on the Environmental and Mechanical Properties of Unalloyed Titanium," *Titanium '92 – Science and Technology*, F. H. Froes and I. Caplan, Eds., Proceedings of the 7th World Titanium Conference, San Diego, CA, June 29–July 2, 1992, pp. 2073–2081.

[14] Meyn, D. A. and Brooks, E. J., "Microstructural Origin of Flutes and Their Use in Distinguishing Striationless Fatigue Cleavage from Stress-Corrosion Cracking in Titanium Alloys," *Fractography and Materials Science, ASTM STP 733*, L. N. Gilbertston and R. D. Zipp, Eds., ASTM International, West Conshohocken, PA, 1981, pp 5–31.

[15] Ward-Close, C. M. and Beevers, C. J., "The Influence of Grain Orientation on the Mode and Rate of Fatigue Crack Growth in α-Titanium," *Metallurgical Transactions A*, Vol. 11A, June 1980, pp. 1007–1017.

[16] Bundy, K. J., Marek, M., and Hochman, R. F., "In Vivo and in Vitro Studies of the Stress–Corrosion Cracking Behavior of Surgical Implant Alloys," *Journal of Biomedical Materials Research*, Vol. 17, 1983, pp. 467–487.

[17] Wang, S.-H. and Muller, C., "A Study on the Change of Fatigue Fracture Mode in Two Titanium Alloys," *Fatigue & Fracture of Engineering Materials & Structures*, Vol. 21, 1998, pp. 1077–1087.

[18] Curtis, R. E., Boyer, R. R., and Williams, J. C., "Relationship Between Composition, Microstructure, and Stress Corrosion Cracking (in Salt Solution) in Titanium Alloys," *Transactions of the ASM*, Vol. 62, 1969, pp. 457–469.

[19] Nakajima, M., Shimizu, T., Kanamori, T., and Tokaji, K., "Fatigue Crack Growth Behaviour of Metallic Biomaterials in a Physiological Environment," *Fatigue & Fracture of Engineering Materials & Structures*, Vol. 21, 1998, pp. 35–45.

[20] Pohler, O. E. M., "Failures of Metallic Orthopedic Implants," *Metals Handbook: Ninth Edition, Manufactured Components and Assemblies*, Vol. 11, pp. 670–694.

[21] Zardiackas, L. and Dillon, L. D., "Failure Analysis of Metallic Orthopedic Devices,' *Encyclopedia Handbook of Biomaterials and Bioengineering – Part B: Applications*, Vol. 1, D. L. Wise, D. J. Trantolo, D. E. Altobelli, M. J. Yaszemski, J. D. Gresser, E. R. Schwartz, Eds., pp. 123–170.

[22] Judy, R. W., Rath, B. B., and Caplan, I. L., "Stress Corrosion Cracking of Pure Titanium as Influenced by Oxigen Content," P. Lacombe, R. Tricot, G. Béranger, Eds., Proceedings – Part IV, Sixth World Conference on Titanium, France, June 6–9, 1988.

[23] Seagle, S. R., Seeley, R. R., and Hall, G. S., "The Influence of Composition and Heat Treatment on the Aqueous-Stress Corrosion of Titanium," *Applications Related Phenomena in Titanium Alloys, ASTM STP 432*, ASTM International, West Conshohocken, PA, 1968 pp. 170–188.

[24] Schutz, R. W., "Stress-Corrosion Cracking of Titanium Alloys," *Stress-Corrosion Cracking – Materials Performance and Evaluation*, Chapter 10, ASM International, Materials Park, Ohio, 1992, pp. 265–278.

[25] *ASM Handbook, "Corrosion," Vol. 13*, Formerly 9th ed., Metals Handbook, ASM International, Materials Park, Ohio, 1945–1973

[26] Gasser, B. and Frenk, A., "Anodizing Medical Titanium Implants," *mo metal surface, Coating of metal and synthetic material*, February, 1997.

[27] Disegi, J. A., "Anodizing Treatment for Titanium Implants," J. D. Bumgardner and A. D Puckett, Eds., *Proceedings of the 16th Southern Biomedical Engineering Conference*, Biloxi MS, April 4–6, 1997, pp. 129–132.

[28] Buhl, H., "Validity of the Slow Straining Test Method in the Stress Corrosion Cracking Research Compared with Conventional Testing Techniques," *Stress Corrosion Cracking— The Slow Strain-Rate Technique, ASTM STP 665*, G. M. Ugiansky and J. H. Payer, Eds. ASTM International, West Conshohocken, PA, 1979, pp. 333–346.

[29] Kim, C. D. and Wilde, B. E., "A Review of the Constant Strain-Rate Stress Corrosion Cracking Test," *Stress Corrosion Cracking—The Slow Strain Rate Technique, ASTM STP* G. M. Ugiansky and J. H. Payer, Eds., ASTM International, West Conshohocken, PA, 1979 pp. 97–112.

[30] *Stress-Corrosion Cracking, Materials Performance and Evaluation*, R. H. Jones, Ed., ASM International, Materials Park, Ohio, 1992.

[31] *ASM Handbook, "Fatigue and Fracture," Vol. 19*, "Stress-Corrosion Cracking and Hydrogen Embrittlement," G.H. Koch, CC Technologies, Inc., ASM International, Materials Park, Ohio, 1996, pp. 483–492.

[32] Cook, S. D., et al., "Fatigue Properties of Carbon-and Porous-Coated Ti-6A1-4V Alloy,' *Journal of Biomedical Materials Research*, Vol. 18, 1984, pp. 497–512.

[33] Boyer, R., Welsch, G., Colings, E. W., "Commercially Pure and Modified Ti," *Material Properties Handbook of Titanium Alloys*, ASM International, S. Lampman, Ed., Materials Park, OH, 1994, pp. 136–237.

Journal of ASTM International, July/August 2005, Vol. 2, No. 7
Paper ID JAI12786
Available online at www.astm.org

Michael D. Roach,[1] *R. Scott Williamson,*[1] *and Lyle D. Zardiackas*[1]

Comparison of the Corrosion Fatigue Characteristics of CP Ti-Grade 4, Ti-6Al-4V ELI, Ti-6Al-7Nb, and Ti-15Mo

ABSTRACT: The purpose of this study was to evaluate and compare the corrosion fatigue (CF) characteristics of a series of titanium alloys including Grade 4 CP Ti (ASTM F 67), Ti-6Al-4V (ASTM F 136), Ti-6Al-7Nb (ASTM F 1295), and Ti-15 Mo (ASTM F 2066). Evaluation of alloy composition, microstructure, and Vicker's microhardness was performed to ensure that each material met the required specification. Smooth and notched CF tensile samples of each alloy were machined using low stress grind techniques and tested at 1Hz according to ASTM F 1801 in both Ringer's solution and distilled/de-ionized water at 37°C. Smooth CF samples had a 10 mm gauge length and a 2.5 mm gauge diameter, and notched CF samples had a 2.5 mm notch root diameter (K_t=3.2). A minimum of three samples was tested at five tension-tension sinusoidal load levels including a run-out level at 10^6 cycles. SEM analysis was performed on the fractured surfaces of representative samples of each alloy to characterize and compare the failure mechanisms.

Fatigue results revealed no differences between the smooth and notched samples of each alloy run in distilled water and those run in Ringer's solution. Results indicated corrosion fatigue mechanisms were not contributing to the fractures under these conditions. However, a significant reduction of fatigue strength was observed for the notched samples of each alloy compared to the smooth samples in identical solutions. These results suggest that even though there is no notch sensitivity in these alloys under static conditions, a notch under dynamic fatigue mechanisms may cause a substantial reduction in cycles to implant fracture. SEM analysis showed typical fatigue fracture morphologies on both the smooth and notched samples. Notch samples, however, did show more defined fluting in terracing due to the higher tri-axial stress state at the root of the notch. In conclusion, corrosion fatigue mechanisms were not contributing to the fracture of these alloys under the given conditions, but the presence of a notch significantly reduced the fatigue strength of all alloys tested.

KEYWORDS: corrosion fatigue, titanium, notch sensitivity, fractography

Introduction

Titanium and its alloys have many favorable properties for implantation into the human body including low weight, high strength, low modulus, superior corrosion resistance, and excellent biocompatibility. It has been estimated that over 2.2 million pounds of titanium devices are implanted into patients worldwide every year [1]. These alloys are known to exhibit good mechanical properties under single cycle loading applications in both smooth and notched configurations. Notch tensile strength to smooth tensile strength ratios for titanium and its alloys have previously been reported to range between 1.4 and 1.7 indicating no effects of notch sensitivity [1,2].

Implants are subjected to discontinuous cyclic loading often over a period of many years. Since the fatigue fracture of these implants is not a function of time but rather stress and number of cycles, the behavior of these alloys under low and high cycle fatigue loading conditions is also

Manuscript accepted for publication 7 January 2005; published July 2005. Presented at ASTM Symposium on Titanium, Niobium, Zirconium, and Tantalum for Medical and Surgical Applications on 9-10 November 2004 in Washington, DC.
[1] Senior Materials Engineer, Senior Materials Engineer, Chair, Department of Biomedical Materials Science, and Professor of Orthopaedic Surgery, University of Mississippi Medical Center, Jackson, MS 39216.

of primary concern. Since the actions of muscles and body weight are the two main forces applied *in vivo*, the load amplitude ranges in these cyclic laboratory tests should remain relatively small [3]. ASTM F 1801 governs the testing of metallic implant materials under uni-axial fatigue conditions and suggests a stress amplitude ratio, R, of 0.053 and testing at 1 Hz to best simulate physiological conditions. Fatigue fracture has been shown to be highly dependent on the alloy microstructure, and up to 90 % of an alloy's fatigue life may be consumed in the crack initiation stage [4]. However, the presence of a surface defect or notch in a material may lead to an increased stress concentration that may accelerate the initiation of a fatigue crack. For this reason, it is also of interest to test notched fatigue samples. Previous notch fatigue tests of titanium alloys using rotating beam test methods have shown that Ti-6Al-4V, while not exhibiting single cycle notch sensitivity, becomes highly notch sensitive under dynamic multi-cycle loading conditions [5]. Commercially pure titanium has also shown reduction in the fatigue strength of notched samples compared to smooth samples in certain fatigue tests, although not to the extent of Ti-6Al-4V [6]. In general, materials become more notch sensitive at lower stresses and a higher number of fatigue cycles [7].

Of all of the available fracture modes, fatigue is one of the most sensitive to its environment. The human body can be a harsh environment even for a metal with such good corrosion resistance as titanium. Many factors including the fluctuating ionic concentration in the saline environment, the body loading mechanisms discussed earlier, initial fluctuations in pH following implantation, and a number of potential tissue reactions can greatly influence the performance of an implant [8]. The addition of these potentially corrosive environmental effects to the previously mentioned fatigue loading mechanisms also warrants the testing of these alloys under corrosion fatigue conditions.

This study examined the smooth and notched sample fatigue characteristics of a series of titanium alloys in distilled/de-ionized water and Ringer's solution at 37°C. Any enhanced effects on the fatigue performance due to the corrosive Ringer's environment and/or the increased state of stress on the notched samples were noted and discussed.

Materials and Methods

Commercially pure (CP) alpha titanium (ASTM F67-grade 4 with 0.38 % O_2) and three titanium alloys including alpha-beta Ti-6Al-4V ELI (ASTM F 136), alpha-beta Ti-6Al-7Nb (ASTM F 1295), and beta Ti-15Mo (ASTM F 2066) were chosen to represent a range of currently available Ti implant alloys. Each material was provided in the form of 8 mm bar stock. In order to verify information supplied on primary certification documents and to establish baseline mechanical properties for fatigue testing, each alloy was characterized for composition, microstructure, and tensile properties. Quantitative compositional evaluation of the major alloying elements of each alloy was determined using an inductively coupled plasma spectrometer (ICP) with a spark attachment for analysis of solid samples (SPECTRO[2] ICP with LISA). The microstructure of alpha CP titanium grade 4 and beta Ti-15Mo were evaluated at 100× magnification according to ASTM E 112 with all samples prepared in duplicate. The alpha/beta titanium alloy microstructures were evaluated at 200× using ETTC-2 microstructure standards for titanium bars with all samples prepared in duplicate [9].

---

[2] SPECTRO Analytical Instruments, Fitchburg, MA.

To establish baseline data for fatigue testing, smooth and notched mechanical testing were performed on each alloy using an MTS[3] servo hydraulic test system. Five smooth and five notched tensile samples of each alloy were machined using low stress grinding to a maximum surface roughness of 16 micro-inches ($R_a \leq 16$) in the gauge. Smooth samples were prepared with a 36 mm gauge length and a 6 mm gauge diameter. Testing was performed at a 0.3 mm/min stroke rate to yield and a 3.0 mm/min stoke rate from yield to failure. Strain was measured using a 25 mm extensometer. The ultimate tensile strength (UTS), 0.2 % yield strength (YS), elastic modulus (MOD), percentage elongation (%El.) to fracture, and reduction of area (%ROA) were calculated. Notched tensile samples were prepared with a K_t factor of 3.2 and a notch root diameter of 6 mm. For many years, the ratio of the notch tensile strength to the smooth tensile strength was considered the most valid method for evaluating the sensitivity to a notch and the toughness of a material [10]. As specified in the ASTM Test Method for Sharp-Notch Tension Testing with Cylindrical Specimens (E 602), it is widely recognized that since the onset of plastic deformation occurs at the yield strength, the ratio of the notch tensile strength to the smooth 0.2 % yield strength may be a more useful predictor of toughness. Notch tensile testing was performed according to ASTM E 602 such that the load rate did not exceed the limit of 690 MPa/min. The notch tensile strength (NTS), the ratio of NTS to UTS, and the ratio of NTS to 0.2 %YS were determined.

Corrosion fatigue testing was performed in tension-tension on smooth samples of each alloy in both aerated distilled/de-ionized water (DI water) and aerated Ringer's solution at 37°C. Fatigue testing followed the guidelines of ASTM F 1801 using the given R ($\sigma_{min}/\sigma_{max}$) value of 0.053. Smooth fatigue samples were prepared using the same low stress grind techniques as the smooth tensile samples previously described with a 10 mm gage length and a 2.5 mm gage diameter (Fig. 1a). Corrosion fatigue was also performed on notch samples of CP Ti - Grade 4, Ti-6Al-4V ELI, and Ti-15Mo in Ringer's solution at 37°C. Notch fatigue samples were prepared with a K_t factor of 3.2 and a 2.5 mm diameter at the root of the notch (Fig. 1b). Samples were mounted vertically to facilitate complete submersion of the sample through the gauge section, and all fixturing in contact with the solution was either teflon or the same alloy as the samples to eliminate any corrosion couples. A tension-tension load was applied as a sinusoidal wave function in load control at 1 Hz. Triplicate samples were tested at each of five stress levels to failure or to a fatigue run-out limit of 10^6 cycles. Data were plotted as stress versus number of cycles to fracture to generate typical S/N curves for each alloy. Fatigue notch factors (K_f) and fatigue notch sensitivities (q) were computed from the S/N curves for each alloy.

$$K_f = \frac{\sigma_{smooth(unnotched)}}{\sigma_{notch}} \qquad (1)$$

$$q = \frac{K_f - 1}{K_t - 1} \qquad (2)$$

The fatigue notch factor (K_f) shown in Eq 1 is the ratio of the fatigue strength of a smooth sample to the fatigue strength of a notched sample at the same number of stress cycles. Since this value will vary according to the position on the S-N curve, it was calculated for high stress (10k cycle) and low stress run-out limit (1M cycle) levels for each alloy. In addition, the fatigue notch sensitivity (q) shown in Eq 2 was calculated for both high and low stress levels.

After the completion of fatigue testing, analysis of the fracture surfaces from at least one fatigue sample of each alloy which fractured at less than 10^5 cycles and one sample that fractured

---

[3] MTS, Eden Prarie, MN.

at greater than 10^5 cycles were examined by scanning electron microscopy (SEM). Comparison was made between alloys to determine the area of crack initiation, fracture morphology during Stage II fatigue crack propagation, and the final fracture mode (Stage III fatigue). Additionally, comparison was made to identify any differences in fracture surface morphology between fatigue in distilled/de-ionized water and Ringer's solution for each alloy. Finally, comparisons were made between the fracture morphologies of the smooth and notched samples of each alloy.

RA = 16 in Gauge

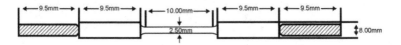

FIG. 1a—Sample drawing for smooth fatigue samples.

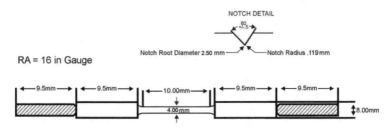

FIG. 1b—Sample drawing for notched fatigue sample.

Results and Discussion

A summary of the smooth and notched mechanical testing results, previously performed in our laboratories, is provided in Table 1 and establishes a baseline for the fatigue analysis [11]. Tensile testing results produced NTS/UTS ratios for each of the alloys that fell within the 1.4–1.7 range defined in previous by others [1,2]. These ratios confirmed that the alloys tested were not notch-sensitive under single cycle mechanical testing conditions. The beta alloy Ti-15Mo showed the least notch sensitivity with a NTS/UTS ratio of 1.53 and a NTS/0.2 %YS ratio of 2.46. A comparison of the smooth sample fatigue curves generated for each alloy is included in Figs. 2a and 2b.

TABLE 1—Smooth and notched mechanical testing results.

Alloy	UTS[a] (MPa)	0.2 % Offset Yield Strength (MPa)	NTS[b] $K_t = 3.2$ (MPa)	NTS/0.2 Yield Strength	NTS/UTS
CP Ti – Grade 4	888	725	1310	1.81	1.47
Ti-6Al-4V ELI	1024	827	1426	1.73	1.39
Ti-6Al-7Nb	978	913	1430	1.57	1.46
Ti-15Mo	874	544	1336	2.46	1.53

[a]UTS=Ultimate Tensile Strength.
[b]NTS=Notch Tensile Strength.

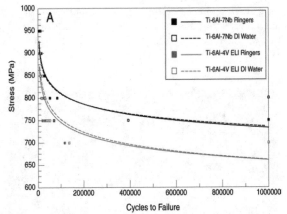

FIG. 2a—*Smooth sample S/N curves generated for Ti-6Al-4V ELI and Ti-6Al-7Nb in Ringers and DI Water.*

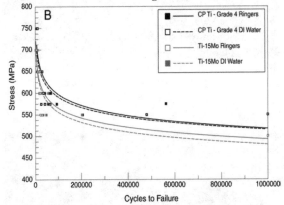

FIG. 2b—*Smooth sample S/N curves generated for CP Ti – Grade 4 and Ti-15Mo in Ringers and DI Water.*

The alpha-structured CP titanium Grade 4 smooth samples produced a fatigue run-out strength of 550 MPa. Other sources have reported similar and slightly lower fatigue strengths for CP titanium Grade 4 for the same number of cycles using rotating beam fatigue techniques [12]. SEM analysis revealed typical ductile fatigue fracture (Fig. 3a) with a crack initiation region showing chevron marks (Fig. 3b). Stage II fatigue on this alloy showed fatigue striations, some secondary cracking, and localized terraces with flutes (Fig. 3c). The terraces and flutes were not unexpected due to the few slip systems available in the hexagonal close packed microstructure of alpha titanium. As the crack propagation continued across the samples, secondary cracking and striation definition were shown to increase to the point of an overload region of failure consisting of ductile dimples (Fig. 3d). Fatigue striations were more finely spaced on samples run to higher cycles (Fig. 4). The same striations and localized terraces with flutes were shown on the DI water samples (Figs. 5 and 6). No signs of corrosion or significant changes in morphology were present on any of the fractured surfaces of this alloy, indicating the chloride concentrations in

Ringers solution did not significantly affect the fatigue properties of this alloy. In addition, no differences were found between the Ringer's and the DI water S/N curves (Fig. 2b) generated for this alloy, which further confirms corrosion assisted fatigue mechanisms were not taking place.

FIG. 3a—*CP Ti – Grade 4 smooth sample fatigue fracture in Ringers solution at low cycles.*

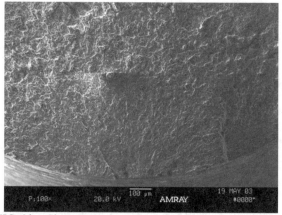

FIG. 3b—*Chevron marks indicating the direction of fatigue crack propagation in Fig. 3a.*

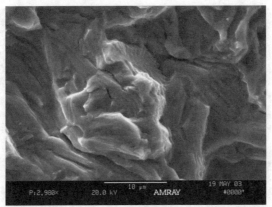

FIG. 3c—*Striations and secondary cracking with localized fluting and terracing in the propagation region of Fig. 3a.*

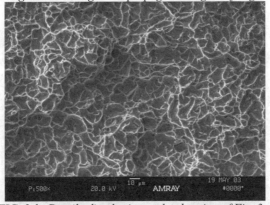

FIG. 3d—*Ductile dimples in overload region of Fig. 3a.*

FIG. 4—*Fatigue crack propagation region of a CP Ti – Grade 4 smooth sample in Ringers at high cycles.*

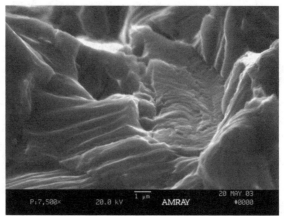

FIG. 5—*Striations, terracing, and fluting in the CP Ti –
Grade 4 smooth sample in DI water at low cycles.*

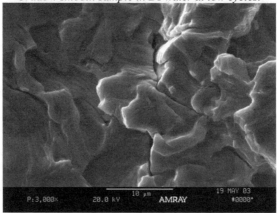

FIG. 6—*Fine striations and secondary cracking on a CP Ti
– Grade 4 smooth sample in DI water at high cycles.*

As expected, the testing of alpha/beta titanium alloys, Ti-6Al-7Nb (TAN) and Ti-6Al-4V ELI (TAV), produced the highest fatigue run-out strengths. The S/N curves for these alloys were very similar, but testing of TAN produced a slightly higher run-out fatigue strength of 750 MPa compared to the 700 MPa of TAV (Fig. 2*a*). The fatigue run-out limit determined for TAV (700 MPa) according to ASTM F 1801 was slightly higher than that determined in previous studies by others (617 MPa) using high frequency rotating beam fatigue tests [4]. This study set the endurance limit at 50 million cycles and ran at 165 Hz compared to testing parameters set by ASTM F 1801 (10^6 cycles at 1 Hz). Once again, no significant differences were shown between the DI water and Ringers solution fatigue curves for these alloys indicating no corrosion assisted fatigue mechanisms were taking place. These results agree well with data from previous studies on fatigue of pure titanium and TAV [13]. Previous studies have also shown that crack growth rates for stress intensity factors above 7 MPa√m are faster for alpha titanium than for alpha/beta

TAV [13]. This is not surprising, as the body centered cubic beta grains in the alpha/beta TAV alloy provide more slip systems and thus fewer constraints on the propagating fatigue crack tip.

SEM fracture analysis of TAV and TAN confirmed these similarities with both alloys exhibiting chevron marks indicative of the area of crack initiation and direction of crack propagation (Figs. 7a and 8a), Stage II fatigue crack propagation with striations and secondary cracking (Figs. 7b and 8b), and Stage III failure region exhibiting ductile overload dimples. The prominent secondary cracking and faint fatigue striations shown in the crack propagation region of TAV agree well with other published studies [14]. Variation in striation orientation (Figs. 7b and 8b) is attributed to changing microstructure and varying grain orientation [15,16]. As expected, more prominent secondary cracking was found on the higher stressed samples, while more closely spaced striations were found on the lower stress samples that failed at higher cycle counts.

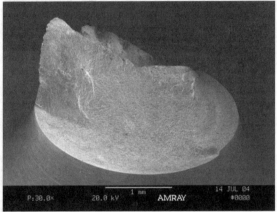

FIG. 7a—*Ti-6Al-4V ELI smooth sample fatigue fracture in Ringers solution at low cycles.*

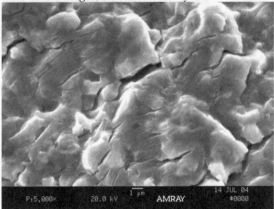

FIG. 7b—*Secondary cracking and fatigue striations in crack propagation region of Fig. 7a.*

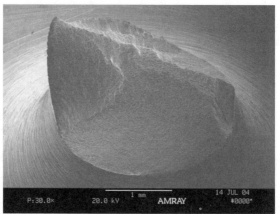

FIG. 8a—*Ti-6Al-4V ELI smooth sample fatigue fracture in Ringers solution at high cycles.*

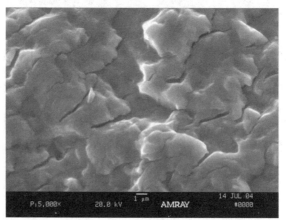

FIG. 8b—*Secondary cracking and fatigue striations in crack propagation region of Fig. 8a.*

As anticipated, due to its lower yield strength, Ti-15Mo had the lowest fatigue strength. SEM analysis revealed more prominent chevron markings (Fig. 9a), with the initial crack propagation region containing striations, furrowing, fluting, and terracing (Fig. 9b). As the crack propagated across the sample, the fatigue mechanism changed to well-defined fatigue striations (Fig. 9c) before ductile overload fracture. On samples run at lower stresses to higher cycles (Fig. 10a), prominent furrowing, and widely spaced striations (Fig. 10b) were shown early in the crack propagation region followed by a transition to finely spaced fatigue striations (Fig. 10c) before an overload failure region on the shear lip (Fig. 10a). It should be noted that substantially less secondary cracking was found on the smooth samples of this alloy compared to the other titanium alloys. No significant differences were shown in the S/N curves in Ringer's solution compared to those in DI water indicating the fatigue mechanisms were not assisted by corrosion.

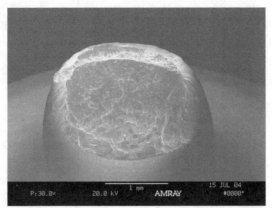

FIG. 9a—*Fracture surface of a Ti-15Mo smooth samplein Ringers solution at low cycles.*

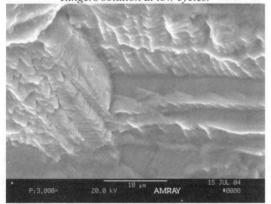

FIG. 9b—*Striations, furrowing, fluting, and terracing in crack propagation region of Fig. 9a.*

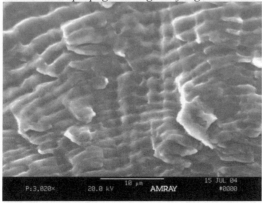

FIG. 9c—*Striations in later crack propagation region of Fig. 9a.*

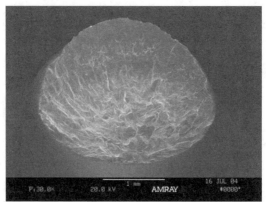

FIG. 10*a*—*Fracture surface of a Ti-15Mo smooth sample in Ringers solution at high cycles.*

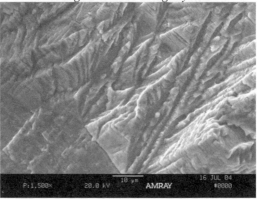

FIG. 10*b*—*Prominent furrowing in the crack propagation region of Fig. 10*a.

FIG. 10*c*—*Striated furrow marks in the crack propagation region of Fig. 10*a.

Since no differences were shown in any of the alloy systems between the S/N curves for Ringer's solution and DI water in the smooth samples, the notched samples were tested in the more aggressive Ringer's solution. A comparison of the notched sample S/N curves generated for each alloy tested in Ringer's solution is included in Fig. 11. A substantial drop in fatigue strength of the notched samples compared to the fatigue strength of the smooth samples was evident in all of the alloys tested. Table 2 provides a summary of the fatigue notch factors, K_f, and fatigue notch sensitivities, q, calculated for each alloy.

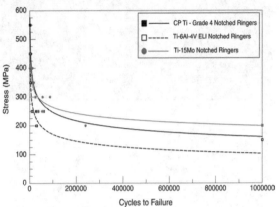

FIG. 11—*S/N curves generated for the notched samples in Ringer's solution.*

TABLE 2—*Fatigue notch sensitivity results, $K_t = 3.2$.*

Alloy	Fatigue Notch Factor, K_f High Stress (10k cycles)	Fatigue Notch Factor, K_f Low Stress (1M cycles)	Fatigue Notch Sensitivity, q High Stress (10k cycles)	Fatigue Notch Sensitivity, q Low Stress (1M cycles)
CP Ti – Grade 4	1.86	2.75	0.39	0.80
Ti-6Al-4V ELI	3.40	4.67	1.09	1.67
Ti-15Mo	1.86	2.50	0.39	0.68

Notched CP Ti Grade 4 samples produced a fatigue run-out strength of 200 MPa, resulting in fatigue notch sensitivity factors on the S/N curve varying between 0.39 at higher stresses and 0.80 at lower stresses. This indicates the material follows the normal trend of becoming more notch sensitive at higher cycle counts and lower stresses. SEM analysis of the high stress samples (Fig. 12a) revealed a fracture pattern consisting of radial markings or ratcheting as the fatigue crack initiated simultaneously from all sides and propagated toward a ductile overload failure region in the center of the fractured surface. Higher magnification in this propagation region showed localized areas of massive fluting, some furrowing, and some striations (Fig. 12b). Analysis of the high cycle samples revealed different fracture morphologies more similar to the smooth samples. One difference compared to the smooth samples was the presence of large continuous relatively flat areas that span the fracture surface to the point of ductile overload (Fig. 13a). These regions showed typical fatigue fracture with fine striations and secondary cracking increasing in frequency as the crack approached the ductile overload area (Fig. 13b).

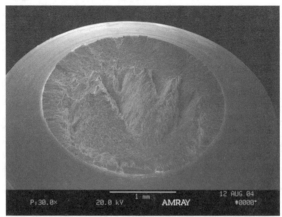

FIG. 12a—*Fatigue fracture of a notched CP Ti-Grade 4 Sample at low cycles.*

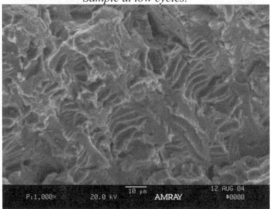

FIG. 12b—*Prominent fluting in the crack propagation region of Fig. 12a.*

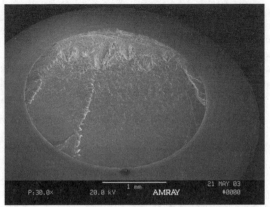

FIG. 13a—*Fatigue fracture of a notched CP Ti-Grade 4 Sample at high cycles.*

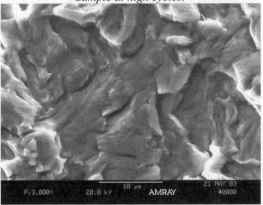

FIG. 13b—*Finely spaced fatigue striations and secondary cracking in the crack propagation region of Fig. 13a.*

In contrast to the higher fatigue strengths produced by the smooth samples, the notched TAV samples revealed the lowest run-out fatigue strength (150 MPa). This large ratio between the run-out strengths for the smooth and notched samples produced the highest notch sensitivity factors (q) ranging from 1.09 for low cycle samples to 1.67 for higher cycle samples, indicating a highly notch sensitive material throughout the fatigue curve. Other studies on this alloy have also shown this notch sensitivity in both axial and rotating beam fatigue configurations [4,17]. SEM analysis revealed a fatigue fracture very similar to the smooth samples (Figs. 14a and 14b). Some ratcheting (Fig. 14a) was noted on the fractured surface and attributed to the higher localized state of stress at the root of the notch.

Beta Ti-15Mo had a lower smooth sample fatigue run-out strength compared to the alpha and alpha/beta alloys. On notched samples, however, it exhibited a fatigue run-out strength of 200 Mpa and notch sensitivity factors equivalent or better than the other titanium alloys. This was expected, considering the single cycle mechanical testing results also proved this material to be the least notch sensitive. Fatigue notch sensitivity values for this alloy ranged from 0.39 at low cycles to 0.68 at higher cycles following the typical pattern of becoming more notch sensitive with increasing cycles. SEM analysis on low cycle samples showed some ratcheting (Fig. 15a), cleavage planes with feather marks, terraces, and substantial fluting and furrowing. As the crack continued to advance across the surface, the mechanism transitioned to a large region of defined striations (Fig. 15b) and eventually to a failure region of ductile overload dimples. Samples run at higher cycles showed a similar fracture surface (Fig. 16a) with finer spacing in the striations (Fig. 16b) indicative of high cycle fatigue.

Conclusions

No significant differences were shown in the S/N curves or fracture analysis for any of the alloys between those run in Ringer's solution and those run in distilled/de-ionized water. This indicates that corrosion processes were not significantly contributing to the fatigue mechanisms under these simulated physiological conditions.

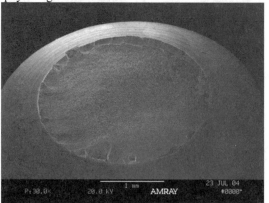

FIG. 14a—*Fatigue fracture of a notched Ti-6Al-4V ELI sample at low cycles.*

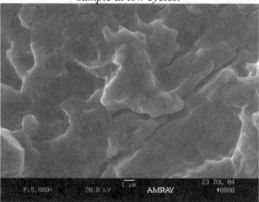

FIG. 14b—*Secondary cracking and fatigue striations in crack propagation region of Fig. 14a.*

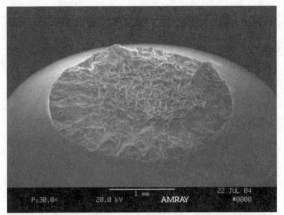

FIG. 15a—*Fatigue fracture of a notched Ti-15Mo sample in Ringers solution at low cycles.*

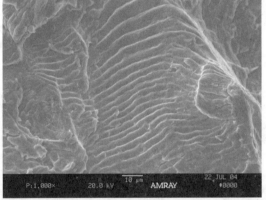

FIG. 15b—*Fatigue striations in crack propagation region of Fig. 15a.*

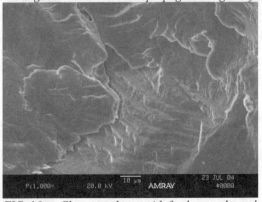

FIG. 16a—*Cleavage planes with feather marks and terracing in Ti-15Mo notched in Ringers at high cycles.*

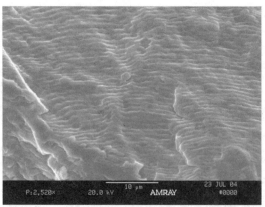

FIG. 16b—*Finely spaced fatigue striations in Ti-15Mo notched samples in Ringers solution at high cycles.*

In addition, while none of the alloys tested showed notch sensitivity under single cycle loading conditions, all of the alloys evaluated showed some degree of notch sensitivity in fatigue. Ti-6Al-4V ELI proved to be highly notch sensitive in fatigue throughout the S/N curve.

Finally, Ti-15Mo exhibited the lowest smooth sample fatigue run-out strength of the alloys tested. However, notched fatigue samples of this alloy were less notch sensitive and generated equivalent or higher fatigue run-out strengths than CP Ti-grade 4 or Ti-6Al-4V ELI.

Acknowledgments

This research was supported by a grant from Synthes. The materials for this research were provided by Synthes and Allvac.

References

[1] Titanium Information Group, Data Sheet No. 14: Titanium for Medial Applications, Issue 1, August 2001.
[2] Zardiackas, L. D., Mitchell, D. W., and Disegi, J. A., "Characterization of Ti-15Mo Beta Titanium for Orthopaedic Implant Applications," *Medical Applications of Titanium and Its Alloys*, S. A. Brown and J. E. Lemons, Eds., ASTM International, West Conshohocken, PA, 1996, pp. 60–75.
[3] Bratina, W. J., et al., "Fatigue Deformation and Fractographic Analysis of Surgical Implants and Implant Materials," *Conference Proceedings of the International Conference and Exhibits on Failure Analysis*, Montreal, Quebec, Canada, 1991, pp. 299–310.
[4] Buck, O., Morris, W. L., and James, M. R., "Fracture and Failure: Analyses, Mechanisms, and Applications," *Proceedings of the ASM Fracture and Failure Sessions at the 1980 Western Metal and Tool Exposition and Conference*, American Society for Metals, P. P. Tung, et al., Eds., Metals Park, OH, 1981.
[5] Cook, S. D., et al., "Fatigue Properties of Carbon- and Porous-Coated Ti-6Al-4V Alloy," *Journal of Biomedical Materials Research*, Vol. 18, 1984, pp. 497–512.
[6] Boyer, R., Welsch, G., Colings, E. W., "Commercially Pure and Modified Ti," *Material Properties Handbook of Titanium Alloys*, ASM International, S. Lampman, Ed., Materials

Park, OH, 1994, pp. 236–237.

[7] ASM Handbook Committee, "Fatigue Failures," *Metals Handbook 8th ed.*, Vol. 10, American Society for Materials, Metals Park, OH.

[8] Dumbleton, J. H. and Edward, H. M., "Failures of Metallic Orthopaedic Implants," *Metals Handbook 8th ed.*, Vol. 10, American Society for Metals, H. E. Boyer, Ed., Metals Park, Ohio, 1975, pp. 571–580.

[9] European Titanium Producers Technical Committee, "Microstructural Standards for α+β Titanium Alloy Bars," Technical Committee of European Titanium Producers, IMI Kynoch Press, England, 1979.

[10] Dieter, G. E., "Mechanical Behavior of Materials Under Tension," *Metals Handbook 9th ed.*, Vol. 8, American Society of Metals, J. R. Davis, Ed., Metals Park, Ohio, 1987, pp. 20–27.

[11] Williamson, R. S., Roach, M. D., and Zardiackas, L. D., "Comparison of Stress Corrosion Cracking of Ti-15Mo, Ti-6Al-7Nb, Ti-6Al-4V, and CP Ti," *Journal of Testing and Evalution*, ASTM International, 2004.

[12] ASM Handbook Committee, "Properties and Selection: Stainless Steels, Tool Materials and Special-Purpose Metals," *Metals Handbook 9th ed.*, Vol. 3, American Society for Materials, Metal Park, OH, 1980, p. 378.

[13] Nakajima, M., Toshihiro, S., Toshitaka, K., and Keiro, T., "Fatigue Crack Growth Behaviour of Metallic Biomaterials in a Physiological Environment," *Fatigue and Fracture of Engineering Materials and Structures*, Vol. 21, 1998, pp. 35–45.

[14] "Failure Analysis of Metallic Materials by Scanning Electron Microscopy," *Iitri Fracture Hanbook,* Metals Research Division IIT Research Institute, Bhattacharyya, S. et al., Eds., Chicago, IL, 1979.

[15] Ward-Close, C. M. and Beevers, C.J., "The Influence of Grain Orientation on the Mode and Rate of Fatigue Crack Growth in α-Titanium," *Metallurgical Transactions A*, Vol. 11A, 1980, pp. 1007–1017.

[16] Wagner, L., "Fatigue and Fracture Properties of Titanium Alloys: Fatigue Life Behavior," *ASM Handbook Vol. 19 Fatigue and Fracture*, ASM International, S. R. Lampman, Ed., Materials Park, OH, 1987.

[17] ASM Handbook Committee, "Properties and Selection of Metals," *Metals Handbook 8th ed. Vol. 1*, American Society of Materials, 1961, p. 530.

Journal of ASTM International, October 2005, Vol. 2, No. 9
Paper ID JAI12785
Available online at www.astm.org

R. Scott Williamson,[1] *Michael D. Roach,*[1] *and Lyle D. Zardiackas*[1]

Comparison of Stress Corrosion Cracking Characteristics of CP Ti, Ti-6Al-7Nb, Ti-6Al-4V, and Ti-15Mo

ABSTRACT: The purpose of this research was to evaluate and compare the stress corrosion cracking (SCC) of Ti-15 Mo (ASTM F2066), Ti-6Al-7Nb (ASTM1295), Ti-6Al-4V ELI (ASTM F136), and Grade 4 CPTi (ASTM F67). Evaluation of alloy composition, microstructure, Vickers microhardness, and tensile properties was performed to determine compliance with the appropriate ASTM specification. For SCC testing, smooth tensile samples with a gauge length of 10 mm and a gauge diameter of 4 mm and notched samples with a notch root diameter of 4 mm (K_t=3.2) were prepared using low stress grinding procedures. Three smooth and three notched samples were tested in distilled H_2O and Ringer's solution at 37°C. Testing was performed using the slow extension rate method at a stroke rate of 10^{-5} mm/s according to the guidelines established in ASTM G129. The ratio of the percent elongation (PER) and reduction of area (ROAR) of smooth and notched samples tested in Ringer's solution and distilled water were evaluated. The fracture surfaces of representative samples were also examined for fracture mode identification using SEM.

Results showed that all alloys complied with the appropriate ASTM specification. SCC may be considered to occur in a material if the ductile properties in an aggressive media are inferior to the ductile properties in a non-aggressive media. Evaluations of the PER and ROAR ratios in smooth and notched samples for each alloy in Ringer's and distilled water showed no indication of SCC failure mechanisms. SEM examination of the fracture surfaces showed no differences in the fracture morphology regardless of the testing solution. These results were consistent with the mechanical testing data. It is therefore concluded that SCC mechanisms were not operating or contributing to the fracture of these alloys under the conditions evaluated.

KEYWORDS: stress corrosion cracking, titanium, notched

Introduction

Historically, there has been much discussion involving stress corrosion cracking and its presence or absence in the failure of titanium alloys. Most in-vitro research has been conducted in one or more of the following environments: 1) halide solutions, 2) high temperature, 3) red fuming nitric acid, 4) methanol, or 5) 3.5 % NaCl solutions [1,2,3,4,5,6,7,8,9,10]. These very aggressive mediums are not accurate in simulating in-vivo conditions which medical grade titanium alloys may experience in the human body. Bundy and others have hypothesized that under certain physiological conditions, certain titanium alloys can experience SCC [2,3].

Several factors can contribute to the occurrence of SCC in titanium. SCC is an environmentally assisted cracking (EAC) type of failure mechanism. Therefore, the environment in which the alloy is tested in the laboratory is critical. As previously stated, most environments in which titanium and its alloys have been tested do not simulate physiological conditions. These

Manuscript received 31 August 2004; accepted for publication 30 March 2005; published October 2005. Presented at ASTM Symposium on Titanium, Niobium, Zirconium, and Tantalum for Medical and Surgical Applications on 9-10 November 2004 in Washington, DC.
[1] Senior Materials Engineer, Senior Materials Engineer, and Professor and Chair of Biomedical Materials Science, respectively, University of Mississippi Medical Center, Jackson, MS 39216.

experiments do provide critical information on what variables in the environment can contribute to SCC. The concentration of anions in the solution is one factor. The majority of the research on SCC in titanium has been done in sodium chloride solutions. Chloride, bromide, and iodine anions may promote SCC in titanium alloys [11]. They are believed to reduce the sustainable load and reduce the threshold for SCC (K_{ISCC}) [7]. Temperature of the solution is also a critical contributing factor. Beta (β) titanium alloys are considered unsusceptible to SCC in ambient conditions. At high temperatures, β titanium alloys have shown SCC mechanisms [8,11].

Metallurgical factors also have an effect on the susceptibility of titanium to SCC. Grain size, grain orientation, volume fraction, alloy composition, and mean free path of alpha (α) phase are some of the important factors [11]. Under certain conditions aluminum concentrations above 5 % and oxygen levels above 0.30 % in the composition are known to be critical factors in contributing to SCC in α and α-β titanium alloys. As aluminum concentration increases, the crack velocity through the α phase increases, and the K_{ISCC} decreases. The increase in aluminum and oxygen levels also restricts slip in the hexagonal close packed α phase and promotes SCC due to dislocation pile up [7,8,11,12]. Metallurgical factors can also increase SCC resistance. Increasing dislocation density by cold working reduces the slip length and therefore increases K_{ISCC} [7].

The purpose of this research was to evaluate α, α-β, and β titanium medical grade alloys' susceptibility to SCC in a non-aggressive medium and an aggressive medium in the smooth and notched conditions at physiological body temperature.

Materials and Methods

The materials evaluated in this research were: α phase Grade 4 CP titanium (CP Ti, ASTM F 67), α-β phase Ti-6Al-7Nb (TAN, ASTM F 1295), α-β phase Ti-6Al-4V ELI (TAV, ASTM F 136), and β phase Ti-15Mo (TiMo, ASTM F 2066). Much Grade 4 CP Ti with 0.38 weight % (w%) O_2 was chosen due to results published in the literature on the SCC susceptibility of CP titanium with >0.30w% O_2. Samples of each alloy were prepared from single lots of 8 mm center-less ground round bar stock supplied by Teledyne Allvac[2] and Synthes.[3] Initial evaluations were performed to establish baseline information and to ensure the materials met the appropriate ASTM specifications. Transverse samples were mounted in bakelite, ground, and polished to a final finish with colloidal silica for metallurgical evaluation. CP titanium samples were etched with a solution containing 10 mL hydrofluoric acid (HF), 30 mL nitric acid (HNO_3), and 50 mL distilled water (H_2O) to evaluate grain size. TAN, TAV, and TiMo samples were etched with Krolls reagent (4 mL HNO_3, 2 mL HF, 94 mL distilled H_2O) for microstructure evaluation. Grain size or structure values were determined in accordance with ASTM E 112 or ETTC-2. The Vicker's microhardness (Hv) of each alloy was measured in three areas of a polished transverse sample prior to etching. Quantitative compositional analysis was performed on a SPECTRO[4] inductively coupled plasma spectrometer (ICP) with a LISA spark attachment to confirm that the alloys met the appropriate ASTM specifications for major alloying elements.

Mechanical testing was performed using a MTS[5] servo-hydraulic 812 testing system. Five smooth and five notched samples were machined from each alloy with a maximum surface

---

[2] Teledyne Allvac, Monroe, NC, USA.
[3] Synthes USA, West Chester, PA, USA.
[4] SPECTRO Analytical Instruments, Fitchburg, MA, USA.
[5] MTS, Eden Prarie, MN, USA.

roughness of 16 micro-inches (Ra = 16) in the gauge or notch section. Smooth samples were prepared with a 6 mm gauge diameter and a 36 mm gauge length as shown in Fig. 1*a*. Notched samples were prepared with a K_t factor of 3.2 and a notch diameter of 6 mm as shown in Fig. 1*b*. Tensile testing was performed at a stroke rate of 0.3 mm/min to yield and 3.00 mm/min to failure. A 25 mm gauge extensometer was used to measure strain. Young's modulus (MOD), ultimate tensile strength (UTS), 0.2 % yield strength (YS), percentage elongation (PE), and reduction of area (ROA) were calculated. The evaluation of the ratio of the notch tensile strength to the smooth tensile strength has been used to predict the sensitivity to a notch and toughness of a material. ASTM E 602 (Test Method for Sharp-Notch Tension Testing with Cylindrical Specimens) specifies that the ratio of the notch tensile strength to the 0.2 % yield strength may be a more useful predictor of notch sensitivity and toughness. Notched tensile testing was performed according to ASTM E 602 such that the load rate did not exceed the limit of 690 Mpa/min. The notch tensile strength (NTS), the ratio of NTS to UTS, and the ratio of NTS to YS were calculated. Scanning electron microscopy (SEM) was used to evaluate the morphology of the fracture surfaces.

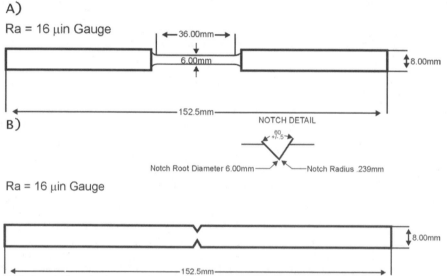

FIG. 1—a) *Sample drawing for smooth tensile samples and* b) *sample drawing for notched tensile samples.*

There are two generally accepted slow strain rate methodologies for determining and evaluating SCC. The first method uses a constant strain rate. In order to perform this test accurately, a strain gauge must be attached to the sample in solution and remain in solution through the duration of the test. This presents a number of potentially significant problems, including retention of the strain gauge on the sample, galvanic coupling of the gauge and the sample, and corrosion of the gauge. The second method is outlined in the ASTM Practice for Slow Strain Rate Testing to Evaluate the Susceptibility of Metallic Materials to Environmentally Assisted Cracking (ASTM G 129). This method incorporates a constant extension rate slow enough that the change in strain rate over the gauge through the duration of the experiment is minimal and will not adversely affect the test results. Because of the potential problems of using the first method, testing was performed according to ASTM G 129. A critical strain rate must be determined to use this test method. Previous research in our laboratories and several investigations by others cited in the *Metals Handbook* has determined that the appropriate strain rate for titanium is in the range of 10^{-5} to $10^{-6}\,s^{-1}$ [9,13,14]

Smooth and notched SCC samples from each alloy were machined from 8 mm round stock. Smooth and notched samples were machined to a R_a of 16 µin. and in the gauge. Smooth samples were prepared with a gauge length of 10 mm and a gauge diameter of 4 mm as shown in Fig. 2*a*. Notched samples are shown in Fig. 2*b* with a 4 mm notch root diameter and a 1 mm notch width to provide a K_t value of 3.2. The SCC mechanical testing was performed on a MTS servo-hydraulic load frame. Triplicate smooth and notched samples were tested in distilled/de-ionized water (DI) and Ringer's solution at 37°C. To obtain the initial critical strain rate of 10^{-6} mm/mm/s the samples were tested at a stroke rate of 10^{-5} mm/s. Additionally, one smooth sample was tested at an initial strain rate of 10^{-5} mm/mm/s and another at 10^{-7} mm/mm/s for each alloy in each solution to validate the critical strain rate. The initial diameter and gauge length for both notch and smooth samples were measured prior to and after testing to an accuracy of 0.001 mm. These measurements were used to calculate PE and ROA. The percentage elongation ratio (PER) and the reduction of area ratio (ROAR) were calculated by dividing the mean values of samples in Ringer's solution by the mean values in distilled/de-ionized water. SEM was used to examine the fracture surfaces.

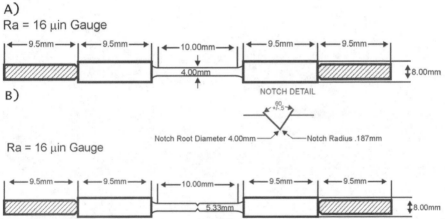

FIG. 2—a) *Sample drawing for smooth SCC samples and* b) *sample drawing for notched SCC samples.*

Results

Microstructure evaluations and Vicker's microhardness values are given in Table 1. The grain size or structure for all alloys tested was found to be acceptable according to ASTM E 112 or ETTC-2. TAN was found to have the highest microhardness with TiMo, TAV, and CPTi following, respectively. ICP compositional analysis results are provided in Table 2. All alloys were found to be within ASTM specifications.

Tensile testing results for smooth and notched samples are summarized in Table 3. TAV samples had the greatest UTS with TAN, CPTi, and TiMo following, respectively. Smooth and notched SCC sample results are given in Tables 4 and 5. Smooth and notched samples for all the alloys tested showed little to no difference in the percentage elongation and the reduction of area values between the aggressive Ringer's solution and the less aggressive distilled/de-ionized water.

TABLE 1—*Microstructure and Vicker's microhardness results.*

Alloy	Microstructure	H_v @ 500 g/12 s
CP Ti Grade 4	ASTM E 112 / 6.5-10.0	269.4 +/- 16.2
Ti-6Al-7Nb	ASTM E TTC-2 / A1	302.9 +/- 10.0
Ti-6Al-4V ELI	ASTM E TTC-2 / A2	295.4 +/- 16.8
Ti-15Mo	ASTM E 112 / 4.0-4.5	296.5 +/- 1.7

TABLE 2—*Compositional analysis of titanium alloys.*

Alloy	Ti	Fe (w%[a])	Al (w%)	Nb (w%)	V (w%)	Mo (w%)	O[b] (w%)
CP Ti Grade 4	Balance	0.04	...	...	...	...	0.38
Ti-6Al-7Nb	Balance	0.10	6.04	7.55	...	...	0.16
Ti-6Al-4V ELI	Balance	0.11	5.92	...	4.25	...	0.11
Ti-15Mo	Balance	0.03	...	...	...	14.55	0.13

[a]w% = weight percent; [b]Value obtained from manufacturer's primary certification.

TABLE 3—*Mechanical testing results of smooth and notched tensile samples.*

Alloy	Smooth Tensile					Notch Tensile	Ratios	
	Elastic Modulus (GPa)	Ultimate Tensile Strength (MPa)	0.2 % Offset Yield Strength (MPa)	Reduction of Area (%)	Percent Elongation (%)	NTS[a] @ K_t = 3.2 (MPa)	NTS/0.2 % Yield Strength (MPa)	NTS / UTS[b]
CP Ti Grade 4	96.7	888.4	725.4	43.9	15.5	1309.7	1.81	1.47
Ti-6Al-7Nb	99.9	978.1	913.4	52.2	16.2	1429.7	1.57	1.46
Ti-6Al-4V ELI	98.4	1023.7	826.8	48.2	14.6	1426.1	1.73	1.39
Ti-15Mo	77.7	873.8	544.3	81.6	20.9	1336.2	2.46	1.53

[a]NTS = Notch Tensile Strength, [b]UTS = Ultimate Tensile Strength.

TABLE 4—*SCC results of smooth samples (n=3).*

Alloy	Solution	% Elongation	% ROA[a]	PER[b]	ROAR[c]
CP Ti Grade 4	DI[d]	15.1 +/- 0.5	41.2 +/- 1.4	0.98	0.98
	Ringer's	14.8 +/- 0.0	40.5 +/- 2.0	...	...
Ti-6Al-7Nb	DI[d]	17.2 +/- 0.2	53.7 +/- 0.7	1.03	0.97
	Ringer's	17.7 +/- 0.5	49.9 +/- 3.4	...	...
Ti-6Al-4V ELI	DI[d]	21.8 +/- 0.9	53.7 +/- 0.7	0.94	0.99
	Ringer's	20.5 +/- 1.2	53.3 +/- 1.0	...	...
Ti-15Mo	DI[d]	29.6 +/- 0.1	78.9 +/- 1.0	0.98	0.99
	Ringer's	29.2 +/- 0.5	78.5 +/- 1.0	...	...

[a]ROA = Reduction of Area; [b]PER = Percentage Elongation Ratio; [c]ROAR = Reduction of Area Ratio; [d]DI = Distilled/De-Ionized Water.

TABLE 5—*SCC results of notched samples (n=3).*

Alloy	Solution	% Notch Elongation	% ROA[a]	PER[b]	ROAR[c]
CP Ti Grade 4	DI[d]	20.0 +/- 1.8	10.2 +/- 0.5	0.97	0.92
	Ringer's	19.3 +/- 0.4	9.3 +/- 0.5	...	...
Ti-6Al-7Nb	DI[d]	17.5 +/- 1.4	7.7 +/- 0.3	1.01	1.00
	Ringer's	17.9 +/- 0.3	7.7 +/- 0.3	...	...
Ti-6Al-4V ELI	DI[d]	20.3 +/- 0.3	9.6 +/- 0.3	1.04	1.03
	Ringer's	21.1 +/- 1.3	9.9 +/- 0.3	...	...
Ti-15Mo	DI[d]	10.5 +/- 0.3	60.5 +/- 2.8	0.99	1.00
	Ringer's	10.4 +/- 0.9	60.7 +/- 2.2	...	...

[a]ROA = Reduction of Area; [b]PER = Percentage Elongation Ratio; [c]ROAR = Reduction of Area Ratio; [d]DI = Distilled/De-Ionized Water.

Discussion and Conclusion

Evaluating the tensile tests showed that TAN had the second highest UTS, but was found to have the greatest YS followed by TAV, CPTi, and TiMo. TAN, TAV, and CPTi were found to have similar ROA and %EL values, which were less than those determined for TiMo. The greater ROA and %El values seen for β TiMo were because of significantly more slip planes in the body centered cubic (BCC) matrix than those present for α and α-β titanium. Testing of the notched samples revealed TAN and TAV had similar NTS values. CPTi and TiMo were also found to have similar NTS values, but less than the other two alloys. The evaluation of the NTS/YS and the NTS/UTS ratios shows that all alloys were significantly greater than 1.00. This indicates that the alloys tested are not notch sensitive with TiMo being the least notch sensitive at a K_t of 3.2. The SEM analysis of the fracture surfaces showed ductile cup and cone type fractures indicative of ductile tensile overload failure.

Materials are defined to be susceptible to stress corrosion cracking if the percentage of elongation and reduction of area in an aggressive corrosion medium are significantly inferior to those properties in a non-aggressive solution. The PER and ROAR ratios are used as indicators to determine if SCC behavior is present. A ratio of less than 0.90 for either indicates that SCC mechanisms may be present [13]. The closer to 1.00 the ratios are, the less susceptible to SCC the material is said to be. The PER and ROAR ratios for all alloys tested were near 1.00. Therefore, SCC mechanisms were not expected to be present on the fracture surfaces. Of note is the ROAR value of notched CPTi. While above the 0.90 criterion, it showed more sensitivity to SCC testing in the notched condition than the other alloys. This is due to the low elongation of the notched samples and the few slip planes in the hexagonal close-packed matrix of CPTi.

SEM analysis of the smooth SCC samples revealed an equiaxed ductile dimple type of fracture for all of the alloys. Representative SEM micrographs of the β titanium (Ti-15Mo) in Ringer's solution showed no evidence of SCC mechanisms (Figs. 3 and 4). For comparison, a representative micrograph of an α-β titanium (TAV) tested in Ringer's solution is also shown (Fig. 5). The morphology and size of the dimples seen in the TAV, TAN, and CP Ti were similar. The β TiMo was found to have much larger dimples on the fracture surface. This is due to the BCC matrix with significantly more slip planes and the larger grain size of the material. The TiMo samples examined also showed slip lines in the larger dimples (Fig. 4). Comparing the fracture surfaces of the smooth SCC to the smooth tensile samples concluded that they were essentially the same size and morphology with minimal differences. Furthermore, there were no differences found between the smooth SCC samples tested at a faster or slower strain rate in either solution.

FIG. 3—*Fracture surface of a Ti-15Mo smooth SCC sample tested in Ringer's solution at lower magnification.*

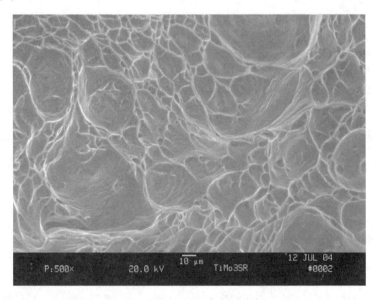

FIG. 4—*Fracture surface of the sample in Fig. 3 at higher magnification.*

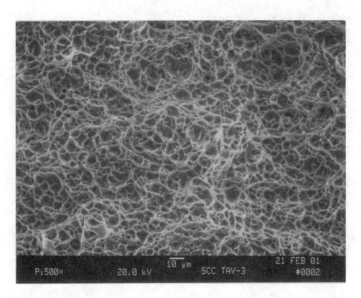

FIG. 5—*Fracture surface of a Ti-6Al-4V ELI smooth SCC sample tested in Ringer's solution at higher magnification.*

SEM micrographs showing the fracture surface of the α titanium (CP Ti) tested in Ringer's solution are given in Figs. 6 and 7. Examination of the fracture surfaces showed a cup and cone type of fracture with dimples evident at higher magnification. The size and morphology of the dimples were similar to the smooth CP Ti samples. Secondary cracking was prominent throughout the fracture surface due to the high tri-axial state of stress at the root of the notch. A representative micrograph of a notched SCC Ti-15Mo sample tested in Ringer's solution is shown in Fig. 8. The fracture surfaces of the notched TiMo samples were similar to the smooth samples and appeared to have larger dimples than α and α-β titanium. Again, this is due to the BCC matrix having significantly more slip planes and a larger grain size. Secondary cracking was not found on the fracture surfaces of the TiMo, TAN, or TAV notched samples. This is due to the hexagonal close-packed α phase CP Ti having fewer slip systems as compared to the BCC β phase in the other alloys [15]. The fracture morphology of all notched SCC samples revealed an equiaxed dimpled morphology due to micro void coalescence. The notched SCC samples fracture surfaces showed no visual differences in morphology from the smooth SCC or tensile samples. In addition, no difference in fracture morphology was observed between testing media.

In conclusion, these results indicate that SCC mechanisms are not occurring in these alloys regardless of the testing solution, sample geometry, or strain rate.

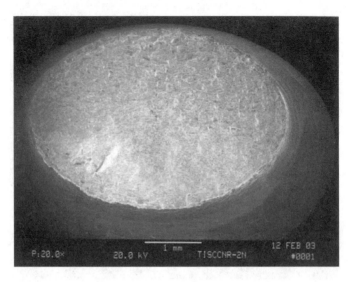

FIG. 6—*Fracture surface of a CP Ti Grade 4 notched SCC sample tested in Ringer's solution at lower magnification.*

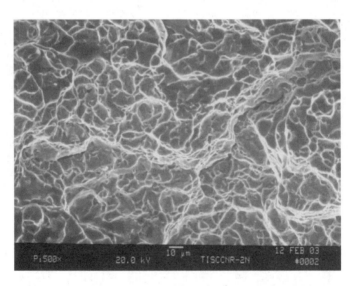

FIG. 7—*Fracture surface of sample in Fig. 6 at higher magnification. Note the secondary cracking.*

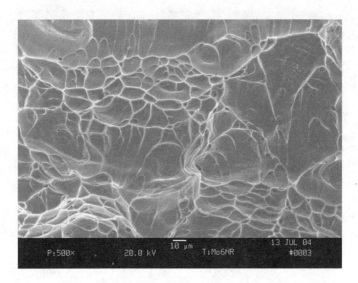

FIG. 8—*Fracture surface of a Ti-15Mo notched SCC sample tested in Ringer's solution at higher magnification.*

References

[1] Judy, R. W., Jr., Caplan, I. L., and Bogar, F. D., "Effects of Oxygen and Iron on the Environmental and Mechanical Properties of Unalloyed Titanium," *Seventh World Titanium Conference*, TMS, F. H. Froes and I. L. Caplan, Eds., Warrendale, PA, 1993, pp. 2074–2080.
[2] Desai, V. H., and Bundy, K. J., "Stress Corrosion Cracking Behavior of Surgical Implant Materials," *Corrosion 84*, National Association of Corrosion Engineers, Houston, TX, 1984, 265, pp. 1–31.
[3] Bundy, K. J., Marek, M., and Hochman, R. F., "In Vivo and In Vitro Studies of the Stress-Corrosion Cracking Behavior of Surgical Implant Alloys," *Journal of Biomedical Materials Research*, Vol. 17, 1983, pp. 467–487.
[4] Hollis, A. C., and Scully, J. C., "The Stress Corrosion Cracking and Hydrogen Embrittlement of Titanium in Methanol-Hydrocchloric Acid Solutions," *Corrosion Science*, Vol. 34, No. 5, 1993, pp. 821–834.
[5] Boyer, R., Welsch, G., and Collings, E. W., *Materials Properties Handbook: Titanium Alloys*, ASM International, Materials Park, OH, 1994, pp. 217–219.
[6] Meyn, D. A., and Sandoz, G., "Fractography and Crystallography of Subcritical Crack Propagation in High Strength Titanium Alloys," *Transactions of the Metallurgical Society of AIME*, Vol. 245, June 1969, pp. 1253–1258.
[7] Curtis, R. E., Boyer, R. R., and Williams, J. C., "Relationship Between Composition, Microstructure, and Stress Corrosion Cracking (in Salt Solution) in Titanium Alloys," *ASM Transactions Quarterly*, ASM International, Materials Park, OH, 1969, pp. 457–469.
[8] Schutz, R. W., "Stress-Corrosion Cracking of Titanium Alloys," *Stress-Corrosion Cracking*, ASM International, R. H. Jones, Ed., Materials Park, OH, 1992, pp. 267–269.
[9] Judy, R. W., Jr., Rath, B. B., and Caplan, I. L., "Stress Corrosion Cracking of Pure Titanium as Influenced by Oxygen Content," *Sixth World Congress on Titanium*, Les Editins de Physique, P. Lacombe, R. Tricot, and G. Beranger, Eds., Les Ulis Cedex, France, 1988, pp.

1747–1752.

[10] Scully, J. C., "Propagation of Stress Corrosion Cracks Under Constant Strain-Rate Applications," *ASTM STP 665*, G. M. Ugiansky and J. H. Payer, Eds., ASTM International, West Conshohocken, PA, 1979, pp. 237–253.

[11] Schutz, R. W. and Thomas, D. E., "Corrosion of Titanium and Titanium Alloys," *ASM Handbook Corrosion Volume 13*, ASM International, Materials Park, OH, 1998, pp. 669–706.

[12] Seagle, S. R., Seely, R. R., and Hall, G. S., "The Influence of Composition and Heat Treatment on the Aqueous-Stress Corrosion of Titanium," *ASTM STP 432*, ASTM International, West Conshohocken, PA, 1968, pp. 170–188.

[13] Zardiackas, L. D., Roach, M. D., Williamson, R. S., and Bogan, J. A., "Comparison of Notch Sensitivity and Stress Corrosion Cracking of a Low-Nickel Stainless Steel to 316L and 22Cr-13Ni-5Mn," *Stainless Steels for Medical and Surgical Applications, ASTM STP 1438*, G. L. Winters and M. J. Nutt, Eds., ASTM International, West Conshohocken, PA, 2003.

[14] Sprowls, D. O., "Evaluation of Stress-Corrosion Cracking," *Metals Handbook*, ASM International, R. H. Jones, Ed., Materials Park, OH, 1987.

[15] Brooks, C. R. and Choudhury, A., *Metallurgical Failure Analysis*, McGraw Hill, Boston, MA, 1993.

BIOLOGICAL AND CLINICAL EVALUATION

Journal of ASTM International, November/December 2005, Vol. 2, No. 10
Paper ID JAI12810
Available online at www.astm.org

Robert M. Urban,[1] *Jeremy L. Gilbert,*[2] *and Joshua J. Jacobs*[1]

Corrosion of Modular Titanium Alloy Stems in Cementless Hip Replacement

ABSTRACT: Severe, localized corrosion of titanium-alloy femoral stems has been reported for specific designs of hip prostheses intended for fixation using acrylic cement. The purpose of the present study was to examine the possibility that corrosion might also occur when titanium-alloy stems are inserted without cement, but the body of the stem is modular in design. Fourteen (7 primary and 7 revision) cementless, modular-body, titanium-6 % aluminum-4 % vanadium alloy femoral stems of similar design were removed at revision surgery after 2 to 108 months *in situ*. The reason for removal was unexplained pain (4), femoral or acetabular loosening (4), infection (3), recurrent dislocation (2), or component malposition (4). The devices and, in selected cases, tissue from the joint psuedocapsule were studied with the use of light and scanning electron microscopy. Fretting corrosion products were characterized using energy dispersive x-ray analysis, selected area diffraction, and micro-Raman spectroscopy. Damage at the modular body connections was absent in 3 stems, mild in 6, moderate in 4, and severe in 1. The surface damage was characterized predominantly by fretting scars and by pitting and etching. Thick deposits of mixed titanium oxides were found adherent to the stem at the sites of corrosion and as 0.01 to 200 micrometer particles within histiocytes and multinucleated giant cells in the joint pseudocapsule. Fretting corrosion at the modular-body junctions of titanium-alloy femoral stems can generate solid degradation products, adding to the particulate burden of the periprosthetic tissues and potentially accelerating bearing-surface wear by a third-body mechanism. Both of these features can potentiate the development and progression of osteolysis. In addition, fretting corrosion can increase the potential for structural failure of the device. These findings stress the importance of the design of modular junctions to minimize corrosion and the generation of corrosion products.

KEYWORDS: titanium, corrosion, fretting, modularity, particulate debris, metal ions

Introduction

Titanium alloy has been employed in several designs of modular-body stems because of its relatively low modulus of elasticity, strength, ability to achieve osseointegration, and its resistance to corrosion [1,2]. Although localized corrosion has been reported at the junction between the neck taper of titanium alloy stems and cobalt-chromium alloy modular heads [3,4], the body of retrieved titanium alloy stems rarely shows evidence of accelerated corrosion, even under conditions of mechanical fretting against bone or cement. Recently, however, severe corrosion has been described at the interface between titanium alloy stems and acrylic cement in specific stem designs [5–7]. This phenomenon is generally not observed in association with stems designed for cementless fixation. Because stem modularity introduces crevice geometries and the potential for fretting corrosion processes, we hypothesized that accelerated corrosion

Manuscript received 2 November 2004; accepted for publication 15 April 2005; published November 2005.
Presented at ASTM Symposium on Titanium, Niobium, Zirconium, and Tantalum for Medical and Surgical Applications on 9-10 November 2004 in Washington, DC.
[1] Assistant Professor and Director, Implant Pathology Laboratory, Professor and Director Section of Biomaterials Research, and Director, Metal Ion Laboratory, respectively, Department of Orthopedic Surgery, Rush University Medical Center, 1653 W. Congress Parkway, Chicago, IL 60612.
[2] Professor and Associate Dean, Department of Biomedical and Chemical Engineering, Syracuse University, Syracuse, NY 13244.

might also occur when titanium-alloy stems are inserted without cement, but the body of the stem is modular in design. The purpose of this study was to examine the mating connections and periprosthetic tissues from 14 modular-body stems of similar design that were removed for cause.

Materials and Methods

Fourteen retrieved, cementless, modular-body femoral stems fabricated from titanium alloy (S-ROM: DePuy, Johnson and Johnson, Warsaw, IN; and Joint Medical Products Corporation, Stamford, CT) were examined for evidence of damage at the mating surfaces of their modular-body connection (Table 1). Fretting and corrosion damage, the composition of associated solid degradation products and, in selected cases, particles of these products in the joint pseudocapsule were characterized.

The modular femoral prosthesis consisted of a titanium-6 % aluminum-4 % vanadium alloy stem that mated with a bead-coated proximal sleeve of the same alloy to form a long, tapered junction. The distal aspect of the stem was fluted and slotted in the coronal plane. The femoral heads were modular and made of cobalt-chromium alloy.

The stems had been implanted in 11 female and 3 male patients with a mean age at implantation of 54 years (range 17 to 83 years). Seven of these patients had received the modular-body femoral stem in a revision operation; and the other seven had received the modular-body implant in a primary operation (Table 1). The femoral components were retrieved at revision surgery 2 to 108 months following implantation. The most common reasons for removal were pain, aseptic loosening, and infection.

TABLE 1—*Clinical data and damage scores for the mating surfaces.*

Implant	Type of Arthroplasty	Duration (Mos.)	Reason for Removal	Surface Damage Scores[a]	
				Stem Taper	Sleeve
1	Primary	5	Dislocation	3	NA
2	Primary	26	Pain	2	2
3	Primary	27	Pain	4	4
4	Primary	49	Pain	2	2
5	Primary	52	Infection	3	3
6	Primary	75	Pain	2	2
7	Primary	108	Aseptic Loosening	3	NA
8	Revision	2	Dislocation	3	3
9	Revision	9	Malposition	1	NA
10	Revision	9	Loosening	1	1
11	Revision	15	Infection	1	1
12	Revision	29	Aseptic Loosening	2	2
13	Revision	34	Infection	2	2
14	Revision	60	Aseptic Loosening	2	2

[a] Surface Damage: (1) None, (2) Mild, (3) Moderate, (4) Severe. NA= not available.

Methods of Analysis

The study material included 14 stems, 11 sleeves (3 had not been removed), and tissue specimens of the joint pseudocapsule associated with Implants 3, 4, 6, 7, 11, and 12 (Table 1). Tissue specimens were not available from the other implants. The retrieved femoral components, degradation products, and tissue specimens were examined using optical and scanning electron microscopes with energy dispersive x-ray analysis (models JSM 5600 and JSM 6460; JEOL, Peabody, MA).

Surface damage to the mating surfaces at the stem and sleeve junction was graded using a stereo light microscope at 8–75× magnification. A corrosion-score ranking system previously described by Gilbert et al. [3] was used to rank the surface damage based on the extent and the severity of fretting and corrosion combined, since these features most often occurred together. The mating surfaces of the stem taper and the sleeve were each graded separately. The degree of surface damage was ranked using a scale of 1 to 4. These scores corresponded to: (1) none: no visible fretting or corrosion, original surface intact; (2) mild: regions of fretting and corrosion confined to one or more small areas, each less than 5 mm in diameter; (3) moderate: markedly larger areas of aggressive local attack at multiple sites with the largest dimension of an individual site less than 20 mm; and (4) severe: large regions of severe surface damage.

Histological sections of the joint pseudocapsules were stained with hematoxylin and eosin and studied by regular and polarized light microscopy. Additional sections were prepared unstained for microanalysis of solid fretting corrosion products. Particles of interest were isolated from the corroded surfaces of the metallic components and from the joint pseudocapsule tissue samples. The isolated particles were analyzed using a laser Raman microprobe spectrometer (Ramascope System 1000; Renishaw, Gloucestershire, United Kingdom) and an analytic electron microscope with selected area diffraction (model JEM-3010, JEOL).

Results

Stem and Sleeve Modular Junction

Fretting and corrosion were observed at the modular-body junction in 11 of the 14 femoral prostheses (Table 1). Surface damage at these connections was ranked as none for 3 implants, mild for 6, moderate for 4, and severe for 1 (Figs. 1a and 1b). Moderate surface damage at the stem and sleeve junction was observed as early as 2 to 5 months following implantation (Implants 1 and 8). The fretting and corrosion damage consisted of discrete areas of fretting scars with pitting and etching of the mating surfaces, often completely obliterating the original surface condition of the implants at these sites. The damage was associated with deposits of adherent black corrosion products. Examination by scanning electron microscopy confirmed that the surface damage was not simply due to mechanical fretting but contained evidence of corrosion as well (Fig. 2).

In each corroded prosthesis for which both components were available, the degree and location of the fretting corrosion damage was similar for the stem and the sleeve. The most frequent sites of damage were at the proximal stem-sleeve junction, but damage also occurred at the distal aspects of the modular connection. Fretting marks were oriented approximately parallel to the long axis of the taper in the majority of the prostheses. In the 6 prostheses with mild surface damage, corrosion and fretting damage was confined to one or more small areas, each less than 5 mm in diameter. The 4 devices with moderate surface damage showed damage at multiple sites with the largest dimension of an individual site ranging from 12 mm to 20 mm (Fig. 1*a*).

Implant 3 was graded as severely damaged. The mating surfaces of this device were extensively damaged and corroded over the majority of the stem (Fig. 1*b*) and the sleeve taper. The surfaces were covered, in part, with abundant, white, or black fretting corrosion products. These products were adherent to the implant surfaces and also consisted of loose particles. There were definite signs of corrosion attack rather than simply fretting. Examination of the stem taper by scanning electron microscopy demonstrated that the small grain microstructure of the alloy was revealed. On the sleeve, the corrosion pattern revealed the alpha grain structure in the alloy as well as pitting.

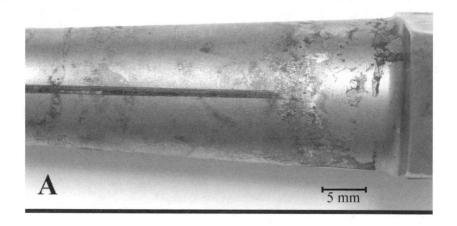

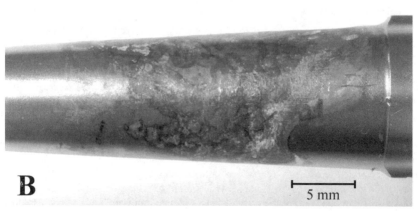

FIG. 1—a) *(Implant 1) The stem and sleeve interface at the proximal aspect of a femoral stem retrieved after 5 months for recurrent dislocation shows moderate fretting and corrosion damage.* b) *(Implant 3) Severe corrosion damage was present on the mating surface of the femoral component of a stem retrieved after 27 months for unexplained pain.*

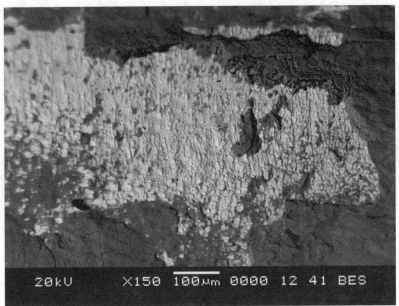

FIG. 2—*(Implant 5) Scanning electron micrograph showing typical fretting corrosion damage at the mating surface of the stem of a modular-body femoral prosthesis retrieved after 52 months for infection and loosening of the acetabular component.*

Particulate Corrosion Products

Energy dispersive x-ray analysis of the solid fretting and corrosion products indicated similar compositions for the deposits adherent to the components compared to the particles in the tissue samples. Particles of titanium oxide that had been generated at the stem and sleeve junctions were found in the joint pseudocapsules associated with Implants 3 and 6. In the stained sections, the particles appeared by light microscopy as amber-colored, refractile, and highly birefringent inclusions within histiocytes and multinucleated giant cells (Figs. 3 and 4). By light and transmission electron microscopy, the titanium oxide particles ranged in size from 0.01 to 200 micrometers. The smallest of these particles had a needle-like appearance (Fig. 5). Electron diffraction and Raman spectroscopy studies (Fig. 6) confirmed that the corrosion products generated at the stem and sleeve junction consisted of polycrystalline mixed oxides of titanium.

Discussion

In this study, surface damage at the tapered junction between the proximal stem and sleeve was found in 11 of 14 modular-body femoral prostheses retrieved at revision. In 6 implants, the degree of damage was mild and probably not clinically important. In another 5 implants, however, the surface damage was considerably more extensive and was graded as moderate or severe. Moderate corrosion damage was found as early as 2 to 5 months following implantation. Although both of these implants had been revised for dislocation, the reason for removal was not believed to be a contributing factor for the observed damage.

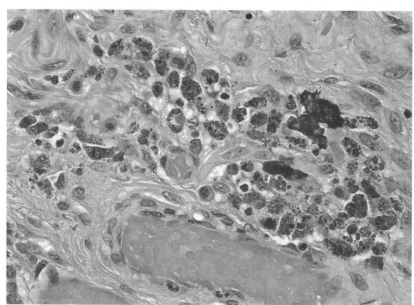

FIG. 3—*Numerous particles, some opaque and others translucent, of titanium oxide fretting corrosion product impart a granular appearance to macrophages in the joint psuedocapsule from Implant 3 (H&E, X156).*

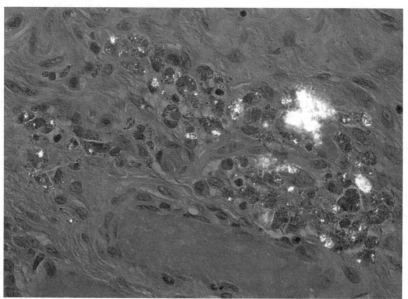

FIG. 4—*The same microscopic field as Fig. 3 under polarized light demonstrates the birefringent nature of the particles of titanium oxide fretting corrosion product.*

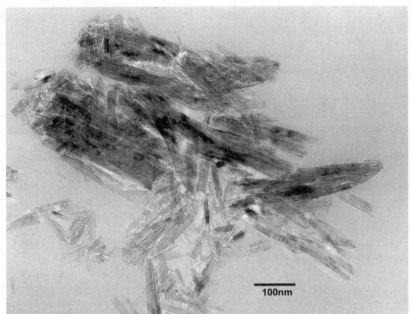

FIG. 5—*Transmission electron microscopy revealed needle-like particles of titanium oxide fretting corrosion product in the joint pseudocapsule tissue of Implant 3.*

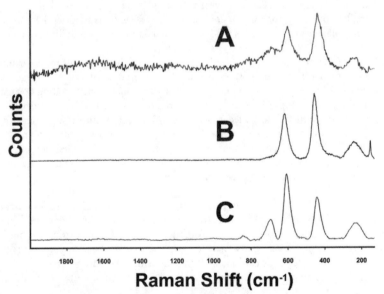

FIG. 6—*Raman spectra of (a) titanium oxide corrosion product from the joint psuedocapsule of Implant 3, (b) titanium oxide rutile standard, and (c) titanium oxide anatase standard.*

Metallurgical examination revealed that the surface damage was caused not only by mechanical fretting, but by crevice corrosion as well. Solid products of these degradation processes, identified as polycrystalline titanium oxides, were deposited at the sites of corrosion. In two implants, one of which demonstrated pronounced titanium metal dissolution (Implant 3), numerous particles of the corrosion products were found in the joint pseudocapsule, suggesting their possible role in a third-body wear process at the bearing surface. The composition of these fretting corrosion products was similar to those described previously in association with crevice corrosion of cemented titanium-alloy femoral stems [7] and distinct from corrosion products such as chromium phosphate previously described in association with corrosion of modular head and neck junctions [8].

The design of a tapered circular conical titanium alloy stem placed within a titanium alloy sleeve, where the sleeve is intended to serve as a bone in-growth site, results in high cyclic loads and stress being imparted across this interface. This configuration presents a substantial potential for fretting motion and creates a large crevice environment into which body fluid can penetrate. The combination of high cyclic stress, fretting, and a large crevice environment can result in conditions permissive for corrosion of titanium alloys.

The mechanism for this corrosion attack involves cyclic loading and fretting motion, which results in abrasion of the oxide films covering the titanium alloy surfaces. The repassivation process for titanium combines metallic titanium with water to yield titanium oxide plus hydrogen ions plus electrons. As the cyclic and repetitive process of depassivation (due to fretting motion) and repassivation occurs, hydrogen ions are generated, and the pH of the crevice solution decreases. When the pH of the solution within the crevice between the stem and sleeve drops to below about 1, the oxide of titanium is unstable and active dissolution of metallic titanium can take place [9,10]. For as long as the pH is at this level, fretting is no longer a necessary condition for continued corrosion. However, continued fretting motion ensures that these low pH conditions remain and that corrosion will continue. The observations reported here are similar to those seen in modular head-neck tapers, and the mechanism of mechanically assisted corrosion described previously was present in these cases as well [11,12].

The findings of the present study stress the importance of the design of modular junctions in order to minimize conditions for corrosion and the generation of corrosion products. It is important to note that while 11 of 14 implants in this study demonstrated fretting corrosion at the body/sleeve junction, all of these implants had been removed for cause. The observations in this study are useful in defining a range of damage that could occur with these devices, but not its prevalence. To determine the true prevalence of the phenomenon, further study of well-functioning implants retrieved postmortem are required. Additionally, much larger, multicenter retrieval studies, including more complete clinical information, could potentially provide data useful for future design improvements in modular-body stems.

In conclusion, modular systems can provide great versatility for the surgeon facing the difficult reconstructive challenges of complex primary or revision total hip replacement [13,14]. However, the surgeon should be aware that fretting corrosion at the modular-body junctions of femoral stems can generate solid degradation products, adding to the particulate burden of the periprosthetic tissues, and potentially accelerating bearing-surface wear by a third-body mechanism. Both of these features can potentiate the development and progression of osteolysis. In addition, fretting corrosion can increase the potential for structural failure of the device [15,16].

Acknowledgments

This study was supported by grant AR39310 from the National Institute of Health. The authors thank Richard Berger, M.D.; Jorge Galante, M.D.; Steven Gitelis, M.D.; Daniel Levin, M.D.; Wayne Paprosky, M.D.; Aaron Rosenberg, M.D.; and Mitchell Sheinkop, M.D. for their cooperation in this study.

References

[1]   Schenk, R., "The Corrosion Properties of Titanium and Titanium Alloys," *Titanium in Medicine. Material Science, Surface Science, Engineering, Biological Responses and Medical Applications*, D. M. Brunette, P. Tengvall, M. Texor, and P. Thomsen, Eds., Springer, Berlin, 2001, pp. 146–70.

[2]   Jacobs, J. J., Gilbert, J. L., and Urban, R. M., "Corrosion of Orthopaedic Implants," *Journal of Bone Joint Surgery*, 1988, Vol. 80-A, pp. 268–82.

[3]   Gilbert, J. L., Buckley, C. A., and Jacobs, J. J., "*In Vivo* Corrosion of Modular Hip Prosthesis Components in Mixed and Similar Metal Combinations. The Effect of Crevice, Stress, Motion, and Alloy Coupling," *Journal of Biomedical Materials Research,* 1993, Vol. 27, pp. 1533–44.

[4]   Goldberg, J. R., Gilbert, J. L., Jacobs, J. J., Bauer, T. W., Paprosky, W., and Leurgans, S., "A Multicenter Retrieval Study of the Taper Interfaces of Modular Hip Prostheses," *Clinical Orthopaedics and Related Research*, 2002, Vol. 401, pp. 149–61.

[5]   Hallam, P., Haddad, F., and Cobb, J, "Pain in the Well-Fixed, Aseptic Titanium Hip Replacement. The Role of Corrosion," *Journal of Bone and Joint Surgery*, 2004, Vol. 86-B, pp. 27–30.

[6]   Scholl, E., Eggli, S., and Ganz, R., "Osteolysis in Cemented Titanium Alloy Hip Prosthesis," *Journal of Arthroplasty*, 2000, Vol. 15, pp. 570–75.

[7]   Willert, H. G., Broback, L.-G., Buchhorn, G. H., Jensen, P. H., Koster, G., Lang, I., et al., "Crevice Corrosion of Cemented Titanium Alloy Stems in Total Hip Replacements," *Clinical Orthopaedics and Related Research*, Vol. 333, 1996, pp. 51–75.

[8]   Urban, R. M., Jacobs, J. J., Gilbert, J. L., and Galante, J. O., "Migration of Corrosion Products from Modular Hip Prostheses. Particle Microanalysis and Histopathological Findings," *Journal of Bone Joint Surgery*, Vol. 76-A, 1994, pp. 1345–59.

[9]   Pourbaix, M., *Atlas of Electrochemical Equilibria in Aqueous Solutions*, Pergamon Press, 1966, p. 218.

[10] Schultz, R. W. and Thomas, D. E., "Corrosion of Titanium and Titanium Alloys," *ASM Handbook Volume 13. Corrosion*, J. R. Davis, Ed., ASM International, Materials Park, OH, 1987, pp. 669–706.

[11] Gilbert, J. L., Buckley, C. A., and Lautenachlager, E. P., "Titanium Oxide Film Fracture and Repassivation: The Effect of Potential, pH and Aeration," *Medical Applications of Titanium and Its Alloys: the Material and Biological Issues, ASTM STP 1272*, S. A. Brown and J. E. Lemons, Eds., ASTM International, West Conshohocken, PA, 1996, pp. 199–215.

[12] Gilbert, J. L. and Jacobs, J. J., "The Mechanical and Electrochemical Processes Associated with Taper Fretting Corrosion: A Review," *Modularity of orthopedic implants, ASTM STP 1301*, J. E. Parr and M. B. Mayor, Eds., ASTM International, West Conshohocken, PA, 1997, pp. 45–59.

[13] Cameron, H. U., "Revision of the Femoral Component: Modularity," *The Adult Hip*, J. J. Callaghan, A. G. Rosenberg, and H. E. Rubash, Eds., Lippincott-Raven, Philadelphia, PA, 1998, pp. 1479–91.

[14] Bono, J. V., McCarthy, C., Lee, J., Carangelo, R. J., and Turner, R. H., "Fixation with a Modular Stem in Revision Total Hip Arthroplasty," *Journal of Bone and Joint Surgery*, 1999, Vol. 81-A, pp. 1326–36.

[15] Heim, C. S., Postak, P. D., and Greenwald, A. S., "Femoral Stem Fatigue Characteristics of Modular Hip Designs," *Modularity of Orthopedic Implants, ASTM 1301*, D. E. Marlowe, J. E. Parr, and M. B. Mayor, Eds., ASTM International, West Conshohocken, PA, 1997, pp. 226–43.

[16] Hoeppner, D. W. and Chandrasekaran, V., "Characterizing the Fretting Fatigue Behavior of TI-6AL-4V in Modular Joints," *Medical Applications of Titanium and Its Alloys: The Material and Biological Issues, ASTM STP 1272*, S. A. Brown and J. E. Lemons, Eds., ASTM International, West Conshohocken, PA, 1996, pp. 252–65.

Journal of ASTM International, July/August 2005, Vol. 2, No. 7
Paper ID JAI12812
Available online at www.astm.org

E.A. Yamokoski,[1] B.W. Buczynski,[1] N. Stojilovic,[2] J.W. Seabolt,[3] L.M. Bloe,[4] R. Foster,[4] N. Zito,[4] M.M. Kory,[5] R.P. Steiner,[6] and R.D. Ramsier[7]

Influence of Exposure Conditions on Bacterial Adhesion to Zirconium Alloys

ABSTRACT: In this paper we combine surface analytical techniques (X-ray photoelectron spectroscopy, X-ray diffraction, infrared reflection absorption spectroscopy, optical and electron microscopy) with viable counts and statistical ANOVA methods to determine the propensity for biological adhesion on zirconium alloy surfaces. We compare the adhesion of laboratory and clinical strains of *Staphylococcus aureus*, *Staphylococcus epidermidis*, and *Pseudomonas aeruginosa* to Zircaloy-2 and Zircadyne-705 materials. Thermal oxidation of the alloys prior to exposure to biological species is also investigated. We present data for 72-h incubation of bacteria and alloys in both shaken and stationary environments. The results of our statistical analysis and experimental observations are relevant to the use of zirconium-based materials for biomedical applications.

KEYWORDS: zirconium alloys, bacterial adhesion, ANOVA, XPS, FTIR, SEM, optical microscopy

Introduction

The use of zirconium and other refractory metals systems for bio-implants is an active area of research. It has been demonstrated recently that oxidized zirconium surfaces exhibit excellent wear-resistant properties [1–3], making them suitable for hip and knee replacements, for example. Our own studies of these materials began with investigating the growth kinetics of thermally grown oxides on these surfaces [4,5], where we initially focused on pure zirconium and Zircaloy-2 (Zry-2) systems. We then studied, in a preliminary manner, the adhesion behavior of several bacteria on Zry-2 and compared this to adhesion on stainless steel surfaces [6]. There, we demonstrated that oxidized Zry-2 did not promote bacterial adhesion and that the presence of bacteria did not cause oxidation or dissolution of the substrates. These preliminary results led us to the conclusion that bio-inert Zry-2 surfaces can be prepared by simple surface processing methods such as polishing and annealing.

The present paper extends our previous work and presents data and analysis concerning the propensity of clinically relevant *Staphylococcus aureus*, *Staphylococcus epidermidis*, and *Pseudomonas aeruginosa* bacteria to adhere to Zircaloy-2 and Zircadyne-705 (Zr705) surfaces. Here we study both laboratory and clinical strains of the bacteria, shaken and stationary bacteriological exposure conditions, and different surface oxide thicknesses. This choice of

Manuscript received 16 August 2004; accepted for publication 28 December 2004; published July 2005.
[1] Graduate Student, Dept. of Biology, The University of Akron, Akron, OH, 44325.
[2] Graduate Student, Depts. of Physics and Chemistry, The University of Akron, Akron, OH, 44325.
[3] Graduate Student, Dept. of Physics, The University of Akron, Akron, OH, 44325.
[4] Undergraduate Student, Dept. of Physics, The University of Akron, Akron, OH, 44325.
[5] Associate Professor, Dept. of Biology, The University of Akron, Akron, OH, 44325.
[6] Associate Professor, Dept. of Statistics, The University of Akron, Akron, OH, 44325.
[7] Associate Professor, Depts. of Physics, Chemistry, and Chemical Engineering, The University of Akron, Akron, OH, 44325-4001, e-mail: rex@uakron.edu.

bacteria includes those found in hospitals and includes both Gram-positive and -negative genera. We use Fourier transform infrared (FTIR) spectroscopy, X-ray photoelectron spectroscopy (XPS), X-ray diffraction (XRD), optical and electron microscopy, viable counts and statistical analysis of variance (ANOVA) methods, and discuss trends that are both experimentally and statistically significant.

Experimental Details

Substrate Preparation

Zircaloy-2 (nominally 1.4 % Sn, Hf depleted, balance Zr) and Zircadyne-705 (nominally 2.5 % Nb, balance Zr + Hf) sheet stock materials (approximately 1.0 mm thick) were received from Wah Chang and cut into nominally rectangular coupons. The sizes and shapes of the coupons varied, depending on the type of analysis for which they were prepared. For IR and viable counts data the coupons were approximately 20 mm × 48 mm, whereas those for XPS and SEM were much smaller. Each substrate was polished on one side using successively finer diamond pastes followed by 0.05 μm alumina suspension. After ultrasonic cleaning and degreasing with acetone, the samples were placed in Petri dishes until needed. These samples are referred to as metallic. For the oxidized samples referred to in this work, a pre-heated convection furnace was used. Polished and cleaned samples from the Petri dishes were placed in the oven for the desired times, removed with the furnace at temperature, and allowed to cool in air at a relative humidity of approximately 50 %. Annealing temperatures were in the range 500–600°C.

Specular reflection FTIR data were collected with a Mattson 7020 spectrometer in the range 700–4000 cm^{-1} at 4 cm^{-1} resolution for various angles of incidence. Following the procedures outlined in our previous work, the resulting interference fringes were analyzed to determine the thickness of the thermally grown oxide layers [4,5]. The IR beam was p-polarized, the entire bench dry-air purged, and a liquid-nitrogen-cooled HgCdTe detector was used. Front surface gold mirrors provided background data needed for obtaining absorbance spectra.

XRD Measurements

Standard XRD measurements in the 2theta geometry were performed on metallic and oxidized substrates, to verify that our oxidation procedure predominantly yields the expected monoclinic form of zirconium oxide on the surfaces. All of the oxide films studied here were in the pre-transition region of the growth kinetics.

Bacterial Adhesion Protocol

Laboratory and clinical strains of *Staphylococcus aureus*, *Staphylococcus epidermidis*, and *Pseudomonas aeruginosa* were incubated overnight at 37°C in tryptic soy broth (TSB). Each bacterium was separately tested in eight flasks, which contained 250 ml of TSB, 0.5 ml of bacterium, and two sample coupons. A four-flask set was incubated at 37°C, and a separate four-flask set was incubated at 37°C on a shaker. The entire eight-flask set was incubated for three days. The broth was carefully removed from each flask after incubation. One metal coupon per flask was aseptically removed and rinsed for one minute in separate 250 ml beakers of sterile 0.85 % (wt/ml) NaCl. Each rinsed coupon was then put into a tube of sterile 0.85 % NaCl and vortexed for one minute, allowed to sit for approximately five minutes, and then vortexed for an additional minute. The second coupon in each flask was aseptically rinsed in a corresponding beaker of saline and placed on a sterile piece of bibulous paper in a sterile glass Petri dish for surface analysis.

Viable Counts

The bacteria in the saline tubes were diluted in sterile 0.85 % NaCl and plated using the spread plate technique. Total dilutions of 2×10^1 to 2×10^7 were used and plated on tryptic soy agar (TSA). The plates were incubated at 37°C overnight. Viable counts were reported in bacteria/ml. Before statistical analysis, these data were corrected for the surface area of the sample coupons, since larger coupons would be expected to yield higher viable counts, with all other factors constant. This correction was performed by first weighing each sample and measuring its thickness. Using mass density values of 6.50 and 6.64 g/cm^3 for Zry-2 and Zr705, respectively, we then calculated the surface area of the samples. This procedure neglects the small area represented by the sides of the coupons and the density of the oxide, but these are minor sources of error. This method does, however, account for any variations in the size and shape of the coupons, since all of them were not cut identically.

Surface Analysis Precautions

It is very important for health and safety reasons that a strict protocol be followed by personnel analyzing samples such as those prepared for this study. We list our protocol here for completeness.

- Do not eat, drink or smoke in the room where work is being performed on bacteria-coated coupons.
- Use forceps, and wear disposable gloves when handling the coupons.
- Leave the coupons in the Petri dishes until the analysis is performed.
- Do not remove the filter paper from the Petri dishes.
- After a coupon has been analyzed, return it to the Petri dish.
- Rinse the surface of the instrument that came into contact with the coupons with Lysol. Allow the Lysol to remain on the surface of the instrument for 5 min, then dry it with a paper towel.
- Keep fingers away from the mouth, nose, eyes, or ears while working with the coupons.
- Thoroughly wash hands with warm/hot water and soap after working with the coupons. The washing should take at least 20 s.
- Clean the forceps after each use by spraying with Lysol and washing with warm/hot water and soap.
- Discard the gloves and used paper towels in an autoclave bag labeled as biohazard for proper removal and disposal by health and safety personnel.

FTIR Characterization

Specular reflection measurements of the adsorbed biofilms are performed with the same IR bench used for the oxide film thickness measurements. For studying the biofilms we used a fixed-angle sampling attachment with gold optics and a polarizer. This provides higher signal-to-noise than the variable angle attachment. In some cases 100 scans or more were averaged as well. For oxidized surfaces, the large intensity interference peaks obscure those from the adsorbed bacteria. However, we do present data below from metallic samples with and without bacteria.

XPS Analysis

The XPS analysis was performed in fixed analyzer transmission mode under high vacuum conditions with pressures often below 8×10^{-9} Torr. We used a Kratos ES-300 electron spectrometer with a dual anode source. For all measurements, the aluminum source was used for the primary survey and detail scans, however the magnesium anode was selected to verify the identity of some Auger features. The X-ray source was operated at 12 kV and 10 mA with the samples approximately 1 cm away from the source. Samples were mounted onto the end of a probe and inserted into a load-lock sample transfer flange for analysis. Argon ion suttering was performed to remove surface contamination and also to further remove material. The sputtering gun was focused and rastered and operated at 2.5 kV. Ultra-high purity argon gas at a total pressure of 2×10^{-5} Torr was used, which corresponds to a sputtering rate of a gold standard of approximately 0.63 nm/min. Depth-profiling values are with respect to this gold standard, and we have not determined differential sputtering rates in this study.

Optical and Electron Microscopy Observations

Microscopic imaging of some of the surfaces was performed with standard optical and SEM techniques. Low-power optical microscopes were used to observe colony formation, and SEM operating at 25 kV was used for identifying bacteria individually. Gold was sputter-coated onto the surfaces to reduce charging effects in SEM measurements.

Results and Discussion

XRD Verification of Substrate Structure

Figure 1 presents representative XRD patterns obtained from the substrates used in this study. Not surprisingly, the strong diffraction peaks from the metallic alpha-Zr phase material (Zr705 in this case) are replaced by monoclinic-ZrO_2 features after thermal oxidation in air [7–9]. The main peaks are from the (-111) and (200) planes and are labeled in the figure with m-ZrO_2 signifying the monoclinic-form of the oxide. There is no evidence for tetragonal-phase zirconia, consistent with what we expect from the literature. All of the oxides layers studied here are in the range before the transition region in the growth kinetics. The thickest oxide layers as determined by IR in this study are 3.1 μm, and all the diffraction patterns that we recorded were similar to those shown in Fig. 1. The two diffraction patterns shown in Fig. 1 have been vertically shifted for clarity of presentation.

Specular IR Reflectance from Surfaces

Specular reflectance IR spectra from metallic Zry-2 surfaces are presented in Fig. 2, where we focus on the region containing the amide modes [10–13]. The lower spectrum is from a coupon exposed for three days to TSB only and serves as a control for our detection of bacterial IR signatures. Both of the spectra in Fig. 2 are from samples exposed to *S. epidermidis* in TSB for the same time period, and clearly show amide II (1555 cm^{-1}) and amide I (1670 cm^{-1}) bands. Note that the intensities of the features from the sample that was shaken during exposure are slightly higher than those from the stationary sample. This observation indicates that more bacteria adhered during shaking and is supported by the ANOVA analysis discussed below. The spectra in Fig. 2 have been vertically shifted for clarity, but no baseline corrections have been made. As mentioned in Section 2, single-pass IR reflection spectroscopy was not able to identify bacteria on oxidized surfaces, since the interference features dominate the spectra [4–6].

XPS Analysis

XPS survey scans are shown in Fig. 3, where the spectra are once again vertically shifted for clarity. Figure 3*a* is from metallic Zr705, before sputter cleaning. The main features present are from zirconium, oxygen, and carbon, as expected. Spectrum 3B is from a similar sample exposed to TSB for three days per the protocol described in Section 2. Note the attenuation of the Zr features and the increase in intensity of the O and C signatures. Traces of phosphorus, nitrogen, and sodium are also evident. This serves as a control for our detection of bacteria by XPS. Finally, spectrum 3C results from a metallic Zr705 surface exposed to *Pseudomonas aeruginosa* in TSB for three days. Note the significant decrease in the Zr features and a further increase in the carbon and sodium signals. Chlorine is also now apparent, and the O(1s) feature is significantly broadened on the high-energy side.

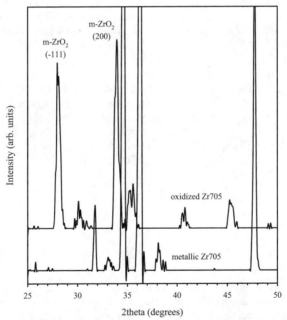

FIG. 1—*X-ray diffraction patterns from metallic and oxidized Zr705, with the latter demonstrating the monoclinic zirconia phase.*

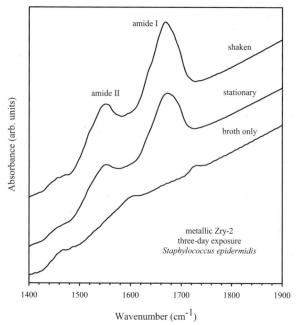

FIG. 2—*Specular reflection IR spectra from metallic Zry-2 surfaces, showing 1555 cm^{-1} amide II and 1670 cm^{-1} amide I bands. Note that these bands are not present from the broth, and that shaken exposure yields slightly more intense bacterial signatures in this case than stationary exposure.*

The decrease in the Zr features due to scattering by adsorbates is a consistent trend that we have observed, similar to the study by Rubio, et al. [13]. The Na and Cl features in our spectra are due to the saline solution used in the sample preparation, and it is clear that much more NaCl resides on surfaces that contain bacteria than on those without biological species. We also observe this in the microscopy studies discussed below. The fact that the C(1s) peak in Fig. 3 also increases with bacterial exposure is consistent with the adhesion of bacteria. Of specific interest here is the O(1s) line, shown in a set of detailed scans in Fig. 4. Here, we see that sputtering removes the high-energy shoulder, which is attributed to oxygen-carbon bonds in the microbial cell surfaces [14]. Sputtering for 30 and 150 min corresponds to the removal of approximately 19 and 95 nm of material, respectively. The latter value is the same order of magnitude as the cell wall thickness of this bacterium, so the observed changes in the spectra are understandable.

Optical Imaging

Optical microscopy proved to be very interesting in this work as well. Many of the samples examined had no observable bacterial growth anywhere on the surface, especially the oxidized coupons. It seemed that metallic surfaces had a higher density of bacteria. The viable counts data did indicate more adhesion on metallic surfaces than on oxidized surfaces, especially for

Pseudomonas aeruginosa, but these observations were not statistically significant. Nevertheless, bacterial adhesion was identified in some cases optically. An example of such a case is given in Fig. 5, which shows *Staphylococcus epidermidis* on an oxidized Zr705 surface in a fractal-like growth pattern.

SEM Studies

Sparse bacterial adhesion on many of the surfaces was verified by SEM imaging. Figure 6 shows one such image, which we present to focus attention on surface defects. This is an oxidized Zr705 surface that was exposed to *Pseudomonas aeruginosa* for three days in a shaken environment. Note that the bacteria have a rod-like appearance, consistent with what has been reported by others [15]. This image and others like it indicate that the bacteria are not preferentially adsorbed in regions of surface defects. This is an interesting result from the standpoint of the effects of surface preparation on bacteriological adhesion to zirconium alloys. We have analyzed other SEM images of oxidized Zr705 surfaces exposed to *Pseudomonas aeruginosa* for up to one month under stationary conditions. Most of these surfaces are void of bacteria, indicating that even long exposures under stationary conditions is not as effective for bacterial growth as are short times with shaking. This observation is consistent with our ANOVA data as well.

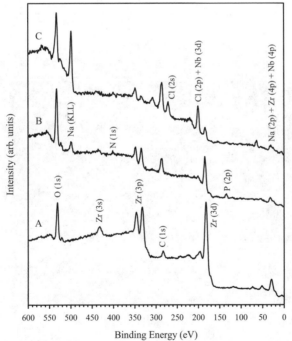

FIG. 3—*X-ray photoelectron survey spectra of metallic Zr705, (A) as polished and cleaned, (B) exposed to broth only, and (C) exposed to Pseudomonas aeruginosa under shaken conditions.*

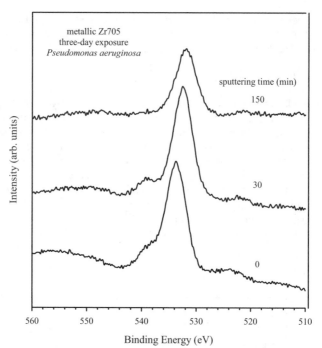

FIG. 4—*The O(1s) region of the upper spectrum of Fig. 3 and the effects of argon ion sputtering.*

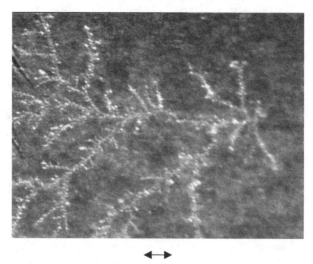

0.1 mm

FIG. 5—*Optical image of a fractal-like colony of Staphylococcus epidermidis on an oxidized Zr705 substrate. The arrow beneath the figure represents 0.1 mm.*

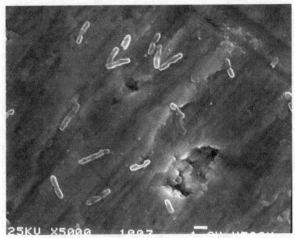

FIG. 6—*SEM image of Pseudomonas aeruginosa bacteria on an oxidized Zr705 substrate after exposure for three days while shaking. There seems to be no enhanced adhesion near defects. The inset scale bar is 1 μm.*

Statistical ANOVA Studies

Adhesion on Zry-2—The data were analyzed using a four-factor ANOVA. The factors were Bacterium (Staphylococcus aureus, Staphylococcus epidermidis, and Pseudomonas aeruginosa), Strain (clinical, laboratory), Condition (shaken, stationary), and Oxidation thickness (0, 1.7, 2.2 μm). The response variable was bacterial growth measured as ln(viable counts/ml/cm^2). The natural logarithm transformation of growth was used to satisfy the ANOVA requirements of normally distributed data with equal variability across factor levels.

First, a model including main effects for each factor and two- and three-factor interactions was tested. It was assumed that there was no interaction simultaneously involving all four factors, thus the mean square for this interaction effect was used as the error for testing the effects in the model. In this analysis, none of the three-factor interactions, nor any effects involving oxidation thickness, contributed meaningfully to the model (P-values ranged from 0.32 to 0.86). Therefore, a second ANOVA model was analyzed that included only main effects for Bacterium, Strain, Condition and their two-factor interactions.

This model (mean square error = 1.63, error df = 25) formed the primary basis of our statistical analysis. In this model, there was a statistically significant Strain x Bacterium interaction (P = 0.005), which indicated that the clinical strain of *P. aeruginosa* exhibited more growth than the lab strain of *P. aeruginosa*. Clinical and laboratory strains of *S. aureus* did not differ significantly from one another, nor did clinical and laboratory strains of *S. epidermidis* differ significantly. Thus, a strain effect existed for *P. aeruginosa* only. The results of this analysis are presented in Fig. 7, in which the data for shaken and stationary conditions are combined.

Another result is that for all bacteria and all strains, there was greater growth under shaken conditions than for stationary conditions (main effect for Condition, P<0.0001). This is demonstrated in Fig. 8. In addition, a Strain x Condition interaction also was found (P = 0.04).

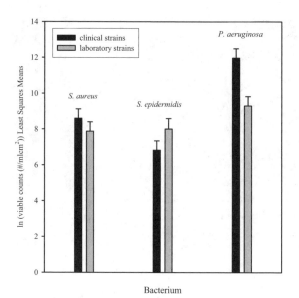

FIG. 7—*Data from ANOVA analysis that indicate a statistically significant difference in the adhesion ability of the three bacteria (*Pseudomonas aeruginosa > Staphylococcus aureus > Staphylococcus epidermidis*) and that in the first case the use of clinical strains enhanced the bacteriological growth.*

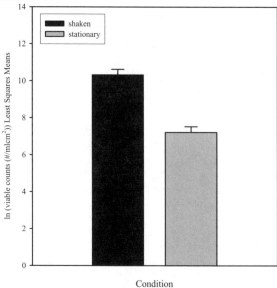

FIG. 8—*Data from ANOVA analysis that indicate a statistically significant difference between stationary and shaken conditions with respect to the adhesion ability of the three bacteria.*

Under shaken conditions, clinical and laboratory strains exhibited very similar growth, but laboratory strains exhibited a greater reduction in growth than clinical strains under stationary conditions. The three species of bacteria also differed in overall mean growth (main effect of Bacterium, P< 0.0001). Specifically, *P. aeruginosa* exhibited more growth than *S. aureus*, which in turn exhibited more growth than *S. epidermidis* (Tukey's HSD multiple comparisons procedure at the 0.05 level of significance). Although not statistically significant (P= 0.28), the observed Bacterium x Condition interaction suggests that *S. aureus* may exhibit a greater reduction in growth under stationary conditions than either *P. aeruginosa* or *S. epidermidis*. The latter two organisms show similar declines (compared to each other) under stationary conditions. These effects are also apparent in Fig. 7.

Adhesion on Zry-2 vs. Zr705—The two materials were analyzed using a four-factor ANOVA. The factors were Bacterium (Staphylococcus aureus, Staphylococcus epidermidis, and Pseudomonas aeruginosa), Strain (clinical, laboratory), Condition (shaken, stationary), and Material (Zry-2 vs. Zr705). All samples analyzed had 2.2 μm oxide thickness. Again, the response variable was bacterial growth measured as ln(viable counts/ml/cm^2). First, a model including main effects for each factor and all two-factor interactions was tested. In this analysis, the Material x Strain (P = 0.73) and Material x Bacterium (P = 0.64) interactions did not contribute meaningfully to the model. So, as before, a second ANOVA model was analyzed with these effects excluded.

The analysis of this model (mean square error = 0.75, error df = 10) led to results for effects *not* involving Material that were similar to those observed in the analysis of Zircaloy-2 alone. Therefore, only the results for effects involving Material are discussed here. Although not statistically significant, we observed less growth on Zr705 than on Zircaloy-2 (main effect of Material, P = 0.06). Also, there was a nearly significant (P = 0.08) Material x Condition interaction, which suggested that both materials exhibit similar growth under stationary conditions. However, while growth is larger on both materials under shaken conditions, the increase in growth is less for Zr705. See Table 1 for a summary of the samples used in this study.

TABLE 1—*Batches of samples exposed to bacteria for three days studied in this work. Here, L stands for laboratory strains and C for clinical strains in parentheses. The abbreviations for the bacteria are SA, SE, and PA, corresponding to Staphylococcus aureus, Staphylococcus epidermidis, and Pseudomonas aeruginosa, respectively.*

Substrate	Oxide Thickness (μm)	Bacterium (strain)	Condition	Viable Counts (#/mlcm^2) × 10^{-3}
Zr705	0.0	PA (L)	shaken	2.45
Zr705	0.0	PA (L)	shaken	2.63
Zr705	1.8	PA (L)	shaken	7.92
Zr705	1.6	PA (L)	shaken	5.16
Zr705	1.7	PA (L)	shaken	6.20
Zr705	2.1	PA (L)	shaken	3.18
Zr705	2.1	PA (L)	shaken	5.58
Zr705	3.1	PA (L)	shaken	7.30
Zr705	2.0	PA (L)	shaken	69.6
Zr705	2.1	PA (L)	shaken	3.62
Zry-2	0.0	SA(L)	stationary	1.12
Zry-2	1.7	SA(L)	stationary	0.24
Zry-2	2.2	SA(L)	stationary	0.07
Zr705	2.2	SA(L)	stationary	...

Substrate	Oxide Thickness (μm)	Bacterium (strain)	Condition	Viable Counts (#/mlcm^2) $\times$ 10^{-3}
Zry-2	0.0	SE(L)	stationary	0.18
Zry-2	1.7	SE(L)	stationary	...
Zry-2	2.2	SE(L)	stationary	0.45
Zr705	2.2	SE(L)	stationary	0.16
Zry-2	0.0	PA(L)	stationary	20.1
Zry-2	1.7	PA(L)	stationary	0.60
Zry-2	2.2	PA(L)	stationary	0.75
Zr705	2.2	PA(L)	stationary	0.55
Zry-2	0.0	SA(L)	shaken	20.6
Zry-2	1.7	SA(L)	shaken	11.5
Zry-2	2.2	SA(L)	shaken	74.6
Zr705	2.2	SA(L)	shaken	9.94
Zry-2	0.0	SE(L)	shaken	24.5
Zry-2	1.7	SE(L)	shaken	56.8
Zry-2	2.2	SE(L)	shaken	11.2
Zr705	2.2	SE(L)	shaken	16.0
Zry-2	0.0	PA(L)	shaken	145
Zry-2	1.7	PA(L)	shaken	5.55
Zry-2	2.2	PA(L)	shaken	241
Zr705	2.2	PA(L)	shaken	10.5
Zry-2	0.0	SA(C)	stationary	1.40
Zry-2	1.7	SA(C)	stationary	0.24
Zry-2	2.2	SA(C)	stationary	2.67
Zr705	2.2	SA(C)	stationary	4.13
Zry-2	0.0	SE(C)	stationary	0.39
Zry-2	1.7	SE(C)	stationary	2.64
Zry-2	2.2	SE(C)	stationary	0.49
Zr705	2.2	SE(C)	stationary	...
Zry-2	0.0	PA(C)	stationary	75.3
Zry-2	1.7	PA(C)	stationary	27.8
Zry-2	2.2	PA(C)	stationary	30.2
Zr705	2.2	PA(C)	stationary	53.7
Zry-2	0.0	SA(C)	shaken	104
Zry-2	1.7	SA(C)	shaken	9.86
Zry-2	2.2	SA(C)	shaken	27.5
Zr705	2.2	SA(C)	shaken	38.2
Zry-2	0.0	SE(C)	shaken	0.86
Zry-2	1.7	SE(C)	shaken	0.35
Zry-2	2.2	SE(C)	shaken	4.16
Zr705	2.2	SE(C)	shaken	0.18
Zry-2	0.0	PA(C)	shaken	467
Zry-2	1.7	PA(C)	shaken	792
Zry-2	2.2	PA(C)	shaken	701
Zr705	2.2	PA(C)	shaken	145

Summary Comments

In this work we tested the ability of clinically relevant *Staphylococcus aureus*, *Staphylococcus epidermidis*, and *Pseudomonas aeruginosa* bacteria to adhere to Zircaloy-2 and Zircadyne-705 surfaces. We demonstrated through the use of viable counts methods that bacterial adhesion to these metal surfaces depends mainly on the exposure conditions (shaken or stationary) and the bacterium of interest. While only significant for *Pseudomonas aeruginosa*, the bacterial strain (clinical origin) also enhances adhesion. To a lesser extent, the specific alloy may make a difference as well. Here we have found no conclusive evidence that surface oxidation enhances or mitigates bacteriological growth. However, it is clear that in the present study, while the samples were being shaken, bacteriological exposures resulted in more growth than for stationary samples. This differs from what we have previously reported on other substrates, however we are using clinical strains in this study. We are currently planning a set of experiments similar to those described here, but with a different collection of clinical strains, to determine if bacterial aggressiveness is related to their original source.

Our XRD data indicate that the surfaces we are studying are either metallic or fully oxidized, and our IR spectra exhibit the amide signatures of adsorbed bacteria in some cases involving metallic substrates. XPS results demonstrate the presence of adsorbed bacteria, especially through the shape and position of the O(1s) feature. Optical microscopy shows that in some cases, macroscopic colonies of bacteria form on these surfaces via fractal-like growth mechanisms. SEM has shown that bacteria do not preferentially adsorb at surface defects, which is another interesting result that needs further investigation. Overall, the data that we have collected thus far indicate that zirconium alloys are relatively inert toward bacterial adhesion and constitute a biocompatible class of materials for use by the biomedical community.

Acknowledgments

We acknowledge support for this effort through NIH-NIBIB grant number EB003397-01, and we are grateful to Wah Chang for providing the Zry-2 and Zr705 materials free-of-charge. We are also appreciative of the assistance provided by Jeannette Killius of the Northeastern Ohio Universities College of Medicine with the SEM phases of this work, and for access to the optical microscope by Professor George Chase.

References

[1] Good, V., Ries, M., Barrack, R. L., Widding, K., Hunter, G., and Heuer, D., "Reduced Wear with Oxidized Zirconium Femoral Heads," *Journal of Bone and Joint Surgery*, American Vol. 85A, 105, Suppl. 4, 2003.

[2] Ries, M. D., Salehi, A., Widding, K., and Hunter, G., "Polyethylene Wear Performance of Oxidized Zirconium and Cobalt-Chromium Knee Components under Abrasive Conditions," *Journal of Bone and Joint Surgery*, American Vol. 84A, 129, Suppl. 2, 2002.

[3] Heuer, D., Harrision, A., Gupta, H., and Hunter, G., "Chemically Textured and Oxidized Zirconium Surfaces for Implant Fixation," *Bioceramics 15 Key Engineering Materials*, 240–2, 789, 2003.

[4] Morgan, J. M., McNatt, J. S., Shepard, M. J., Farkas, N., and Ramsier, R. D., "Optical and Structural Studies of Films Grown Thermally on Zirconium Surfaces," *Journal of Applied Physics 91*, 9375, 2002.

[5] McNatt, J. S., Shepard, M. J., Farkas, N., Morgan, J. M., and Ramsier, R. D., "Non-Destructive Characterization of Films Grown on Zircaloy-2 by Annealing in Air," *Journal of Physics D*, Applied Physics 35, 1855, 2002.

[6] Buczynski, B. W., Kory, M. M., Steiner, R. P., Kittinger, T. A., and Ramsier, R. D., "Bacterial Adhesion to Zirconium Surfaces," *Colloids and Surfaces B: Biointerfaces 30*, 167 2003.

[7] Baek, J. H. and Jeong, Y. H., "Depletion of Fe and Cr within Precipitates During Zircaloy-4 Oxidation," *Journal of Nuclear Materials 304*, 107, 2002.

[8] Kim, H. G, Kim, T. H. ,and Jeong, Y. H., "Oxidation Characteristics of Basal (0002) Plane and Prism (11-20) Plane in HCP Zr," *Journal of Nuclear Materials 306*, 44, 2002.

[9] Huy, L. D., Laffez, P., Daniel, Ph., Jouanneaux, A., Khoi, N. T., and Simeone, D., "Structure and Phase Component of ZrO_2 Thin Films Studied by Raman Spectroscopy and X-ray Diffraction," *Materials Science and Engineering B 104*, 163, 2003.

[10] Ede, S. M., Hafner, L. M., and Fredericks, P. M., "Structural Changes in the Cells of Some Bacteria During Population Growth: A Fourier Transform Infrared-Attenuated Total Reflectance Study," *Applied Spectroscopy 58*, 317, 2004.

[11] Reiter, G., Siam, M., Falkenhagen, D., Gollneritsch, W., Baurecht, D., and Fringeli, U. P., "Interaction of a Bacterial Endotoxin with Different Surfaces Investigated by in Situ Fourier Transform Infrared Attenuated Total Reflection Spectroscopy," *Langmuir 18*, 5761, 2002.

[12] Filip, Z. and Hermann, S., "An Attempt to Differentiate *Pseudomonas* spp. and Other Soil Bacteria by FT-IR Spectroscopy," *Eur. J. Soil Biol. 37*, 137, 2001.

[13] Rubio, C., Costa, D., Bellon-Fontaine, M. N., Relkin, P., Pradier, C. M., and Marcus, P., "Characterization of Bovine Serum Albumin Adsorption on Chromium and AISI 304 Stainless Steel, Consequences for the *Pseudomonas fragi* K1 Adhesion," *Colloids and Surfaces B: Biointerfaces 24*, 193, 2002.

[14] Van der Mei, H. C., de Vries, J., and Busscher, H. J., "X-Ray Photoelectron Spectroscopy for the Study of Microbial Cell Surfaces," *Surface Science Reports 39*, 1, 2000.

[15] Reid, G., Busscher, H. J., Sharma, S., Mittelman, M. W., and McIntyre, S., "Surface Properties of Catheters, Stents and Bacteria Associated with Urinary Tract Infections," *Surface Science Reports 21*, 251, 1995.

Journal of ASTM International, July/August 2005, Vol. 2, No. 7
Paper ID JAI12814
Available online at www.astm.org

Christoph M. Sprecher,[1] *Joachim Kunze,*[2,3] *Björn Burian,*[1] *Nadine Villinger,*[1] *Joshua J. Jacobs,*[3] *Erich Schneider,*[1] *and Markus A. Wimmer*[3]

A Methodology to Fabricate Titanium and Stainless Steel Wear Debris for Experimental Use: A Comparison of Size, Shape, and Chemistry

ABSTRACT: It is well established that particulate debris can cause osteolysis. The current paper describes a simple procedure to generate titanium and stainless steel particulates for in vitro and in vivo use.

The 'wear generator' consisted of three pins out of titanium or stainless steel fixed onto a stir "bar." The bar was rotated in a beaker filled with Ringer's solution against a disk of the same metal. The extracted particles were described using scanning electron microscopy, X-ray diffraction analyses, and inductively coupled plasma - optical emission spectroscopy. In addition, the chemical stability of the generated particles was tested.

The Equivalent Circle Diameter of stainless steel particles was smaller than that of titanium. Titanium particles contained pure titanium metal and titanium oxides. In contrast, stainless steel particles contained soluble corrosion products (e.g., nickel) and other elements from the salty lubricant.

KEYWORDS: wear, particle characterization, titanium, stainless steel, hip prostheses

Introduction

The operative stabilization of fractured bones using plates and screws and the replacement of arthritic joints with prosthetic devices are standard procedures in orthopedic surgery nowadays. However, wear and the consequences of wear are still limiting factors despite the fact that total hip arthroplasty survival rates are at over 90 % after 15 years in-vivo [1]. Submicron particulate debris, generated during wear, has been reported to cause local tissue irritation [2,3] and, in close proximity of bone, osteolysis and subsequent loosening of the implant [4]. The articulating surfaces of prostheses (e.g., ball and cup of a hip implant) are typically the major sources of wear debris, while in several studies it has been reported that also non-articulating surfaces undergo wear damage. Fretting corrosion of taper junctions of modular hip implant systems is a typical example in this context [5]. Also, devices for fracture treatment have been shown to cause osteolysis [6]. In-vitro studies have shown that particle morphology and particle amount influence cell and tissue reactions [7]. Particles of less than 1 μm are typical in the surrounding tissue of prostheses and are of particular importance because they are in the phagocytosable range [9]. Metal debris is typically even an order smaller in size and well below 1 μm [8].

Manuscript received 13 August 2004; accepted for publication 25 January 2005; published July 2005. Presented at ASTM Symposium on Titanium, Niobium, Zirconium, and Tantalum for Medical and Surgical Applications on 9-10 November 2004 in Washington, DC.

[1] AO Research Institute, Clavadelerstrasse, 7270 Davos, Switzerland.
[2] Technical University Hamburg-Harburg, 21071 Hamburg, Germany.
[3] Rush University Medical Center, 1653 W. Congress Pkwy, Chicago, IL 60612, USA.

Particles of the above size range are not readily available for experimental use. Therefore, in the present paper, we describe an economic methodology to generate submicron particulates with titanium and stainless steel implant materials. A detailed characterization of the size, shape, and chemistry of the generated particulates is presented.

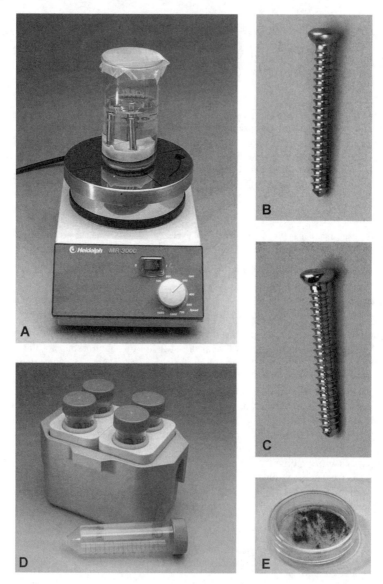

FIG. 1—'Wear generator': magnetic stirrer (A), Ti screw (B), SS screw (SS), 50 ml centrifuge tubes (D), and collected wear particles (E).

Materials and Methods

Two different implant materials were used: (1) commercially pure titanium (Ti) and (2) stainless steel for surgical implants (SS). Titanium was from Grade 4, and the composition was consistent with ASTM Standard F 67. The composition of stainless steel followed ASTM Standard F 138.

The 'wear generator' consisted of a conventional laboratory stirrer with a beaker and discoidal stir "bar," coated with Teflon. Three pins of titanium or stainless steel materials were placed on the underside of the magnetic Teflon stirrer and "articulated" against a metal disk of identical material. Both discoidal stirrer and disk were placed in a beaker and operated using Ringer's solution as the lubricant (Fig. 1). The metal disc had a diameter of 60 mm (regardless of material), and the thickness was 2 mm. Fretting wear particles were generated due to the wobbling motion characteristics of the following Teflon head (hydrodynamic effects) causing stick-slip movements between pins and disk. The particles were extracted from the media by means of ultra-centrifugation, decantation, and accumulation. The slag obtained was dried at 50°C and pulverized using a mortar afterwards. The particles were heat-sterilized for 3 h at 180°C and filled into a ø2.2 mm tube and locked with a piston at one end. After slight compaction, the height of the titanium and steel particles inside the tube was adjusted to 0.5 mm each, thus controlling for equal wear volume for later application in animal experiments. In addition, the weight was controlled using a balance with a nominal precision of 0.01 mg (AT261 DeltaRange, Mettler Toledo, Greifensee, Switzerland). After filling, the other end of the tube was also closed, and the whole device was gamma-sterilized after packaging.

For size and shape analysis, the particles were suspended in ethanol and put on a stub with a double-side carbon tape. The particles were examined using a low-voltage scanning electron microscope without coating (SEM, Hitachi FESEM S-4100, Kyoto, Japan). Images were magnified up to 10 000× in the secondary electron mode at 5 kV acceleration voltages.

Particle size and shape description was performed according to a methodology reported earlier [10]. For this purpose, the area and perimeter of more than 400 particles from both materials were measured. Further, to study frequency distribution, the particles were classified according to their shape-independent size, i.e., according to the so-called 'Equivalent Circle Diameter' (ECD). For titanium, 0.025 μm classes were set up; for example the class '0.200 μm' contained particles with $0.175 \text{ μm} \leq \text{ECD} < 0.200 \text{ μm}$. Due to the smaller particle size, stainless steel classes were smaller, namely 0.0125 μm. For shape characterization, the 'Equivalent Shape Ratio' (ESR) [10], particle volume (V_{ESR}), and surface (S_{ESR}) were calculated.

The elemental composition of the particles was investigated using an Inductively Coupled Optical Emission Spectroscopy System (ICP-OES, Perkin Elmer Optima 2000 DV). Any existing chlorides (and/or other anions) were analyzed using an Ion Chromatography System (IC, Dionex). SS particles were dissolved in a mixture of HCl and HNO_3, and the relative elemental composition was compared to a similarly treated stainless steel following ASTM Standard F 138. To study the soluble fraction (corrosion products) of both titanium and stainless steel wear particles, an aliquot of the particles was extracted with water, centrifuged, and analyzed for Cr^{+3}, Ni^{+2}, Mo^{+5}, Mn^{+2}, Cu^{+2}, Na^{+1}, K^{+1}, Ca^{+2}, and chloride Cl^{-1}. Na, K, Ca, and Cl are elements of the surrounding medium (Ringer's solution) and are introduced into the debris during the isolation process of the wear particles. Applied parameters for ICP-OES and IC are given in the Tables 1 and 2.

To gain more insight into occurring compound structures of generated debris, X-ray diffraction analyses were performed using a D5000 Theta-Theta diffractometer (Siemens,

Germany) equipped with a copper tube. Heat sterilized particles were fixed on glass using hair spray and analyzed at 40 kV and 40 mA in the range of 2–100° 2-Theta. The diffractometer readings were corrected for noise, which was generated from the glass specimen holder.

TABLE 1—*Wavelength and calibration for ICP-OES measurements. The following measuring conditions were used; Sample uptake rate: 1.5 mL/min; Plasma gas: 15 L/min; Auxiliary gas: 0.4 L/min; Nebulizer gas: 0.75 L/min; RF power: 1300 W.*

Element	wavelength [nm]	calibration Std. [µg/mL]	Linearity	detection limit [µg/mL]
Cr	267	20/10/5/1/0.2/0.1	0.99997	0.61
Ni	231	20/10/5/1/0.2/0.1	0.99997	0.67
Mo	202	5/2/1/0.5/0.2/0.1	0.99999	0.28
Mn	257	5/2/1/0.5/0.2/0.1	0.99999	0.07
Cu	327	5/2/1/0.5/0.2/0.1	0.99998	0.20
Fe	238	10/5/1/0.5/0.2/0.1	0.99999	0.16
Na	589	20/10/5/1/0.5/0.2	0.99995	0.25
Ca	317	10/5/1/0.5/0.2/0.1	0.99999	0.12
K	766	20/10/5/1/0.5/0.2	0.99998	0.56
Rh	343	1	internal	standard

TABLE 2—*Measuring conditions for IC.*

HPLC- pump	M590 (Waters)
autosampler	201/431 (Gilson),
conductivity detector	CD 20 with suppressor ASRS-I (Dionex)
Chromatography	was done by a AS4A SC 250*4 mm column connected with a guardian column AG4A 40*4 mm
elution	with a flowrate of 1,5 mL/min
Eluent	1,8 mMol/L sodiumcarbonate and 1,7 mMol/L sodium hydrogencarbonate
centrifugation	16000*g
	20 µL were injected

Results

Both groups of wear particles were comparable in shape but varied in size (Fig. 2). The ECD of the Ti particles was 0.4972 ± 0.4092 µm and thus larger than the SS particles with a diameter of 0.1218 ± 0.0647 μm (Table 3). Approximately 95 % of the SS particles were smaller than 0.5 μm, compared to 63 % of the Ti particles (Figs. 3 and 4). As shown in Fig. 3, 83 % of the Ti particles were smaller than 1.0, and 95 % were smaller than 2.0 μm. The ESR of both materials was 0.3058 ± 0.1141 for Ti and 0.2808 ± 0.1227 for SS. Hence, shape-wise particles were similar. The average particle volume of SS was much smaller compared to Ti (0.0009 versus 0.1257 μm^3). Similarly, the average surface area of SS particles was smaller than that of Ti (0.0494 versus 1.0788 μm^2, Table 3).

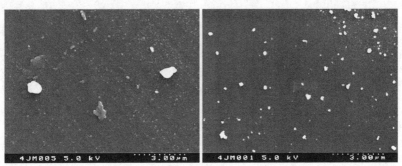

FIG. 2—*REM images of wear particles: titanium (left) and stainless steel (right).*

TABLE 3—*Numerical results of the wear particle characterization made from stainless steel (SS) and titanium (Ti), respectively. Abbreviation: Equivalent Circle Diameter (ECD), Standard Deviation (SD), Equivalent Shape Ratio (ESR), Volume Calculated based on ESR (V_{ESR}), Surface calculated on ESR (S_{ESR}).*

		Ti (ECD < 2.0 μm)	SS (ECD < 0.5 μm)
Equivalent Circle Diameter	ECD ± SD [μm]	0.4972 ± 0.4092	0.1218 ± 0.0647
	(Range)	(0.1089 – 1.9763)	(0.0485 – 0.4428)
Equivalent Shape Ratio	ESR ± SD [-]	0.3058 ± 0.1141	0.2808 ± 0.1227
	(Range)	(0.0334 – 0.7328)	(0.0309 – 0.6704)
Volume calculated based on ESR	$V_{ESR} \pm$ SD [μm^3]	0.1257 ± 0.3144	0.0009 ± 0.0019
	(Range)	(0.0003 – 2.0662)	(0.000018 – 0.022549)
Surface calculated based on ESR	$S_{ESR} \pm$ SD [μm^2]	1.0788 ± 1.8626	0.0494 ± 0.0639
	(Range)	(0.0310 – 10.1645)	(0.0059 – 0.4959)
	Particle No	444	714
	Particle [%]	94.27	95.07

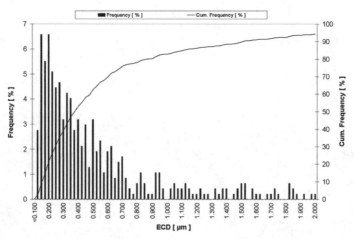

FIG. 3—*Histogram of wear particles: titanium.*

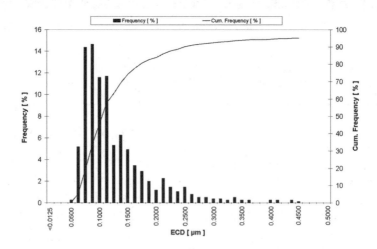

FIG. 4—*Histogram of wear particles: stainless steel.*

Diffraction patterns for Ti displayed mostly metallic titanium and secondary titanium oxides with changing stochiometry. Besides the reflection peak for stainless steel, SS showed further reflections for chromium-nickel-oxides. For both materials, Ti and SS, the crystal peaks were superimposed by a considerable amorphous fraction.

Wear and bulk material differed with regard to their chemical composition (Table 4). Next to the oxidation of particles due to chemical attack (e.g., galvanic corrosion) and tribochemical processes (e.g., local he ating due to friction), the influence of elements of the lubricant became evident (Table 5).

TABLE 4—*Composition from the main elements of the stainless steel (SS) wear particles compared to the ASTM Standard F 138 of stainless steel investigated by ICP-OES.*

	Cr [%]	Ni [%]	Mo [%]	Mn [%]	Cu [%]	Fe [%]
Wear	17.25	10.50	1.00	0.91	0.05	55.10
Bulk	17.66	14.29	2.81	1.74	0.04	~ 64

TABLE 5—*Composition of the corrosion products from commercial pure titanium (Ti) and stainless steel (SS) wear particles investigated by ICP-OES and IC.*

	Cr [mg/Kg]	Ni [mg/Kg]	Cu [mg/Kg]	Na [mg/Kg]	K [mg/Kg]	Ca [mg/Kg]	Fe⁻ [mg/Kg]	Cl⁻ [mg/Kg]
SS	< 15	854	< 5	1039	316	1479	1.85	5400
Ti	< 15	< 15	< 5	150	< 50	125	< 5	< 25

The wear particles showed significant constituents of "non-prosthetic" origin, such as Na, K, Ca, and Cl. In particular for stainless steel, the overall composition seems to differ when compared to the ASTM Standard F 138. As the solubility of chromium oxides in water is low, no chromium was detected in the water eluate of the wear particles. Similarly, titanium was not found because of its low solubility. Nickel ions, however, were present. Interestingly, for the titanium debris, the environmental "contamination" was greatly reduced. Thus, only sodium and calcium originating from the Ringer's solution were picked up. Data are presented in Tables 4 and 5.

Discussion

The presented method was capable in generating submicron wear particles similar to what is found in surrounding tissue of hip prosthesis or other implant devices undergoing wear. The investigated size range and the shape of the particles are comparable to the description in others [3,8,11–13]. Also, it has been shown by others that wear particles in the 0.1–1.0 μm size range are the most biologically active [14]. With approximately 83 % of titanium particles and nearly 100 % of stainless steel particles in that size range, we think we identified an economic way in producing particulate debris for various in vitro and in vivo experiments.

The majority of the wear particles was covered with an oxide layer, but was not completely oxidized. It is suggested that in addition to the oxide layer, particles from the underlying bulk material have been created through various tribological processes. Due to the highly corrosive environment, these particles may repassivate quickly. Our observations correlate well with findings of in vitro tests and retrieved implants by Cook et al. [11]. The chemical composition of wear particles contained mainly elements of the base material and different oxides [15].

Interestingly, the data of this study suggest that the composition of wear particles can differ from bulk. A deviation in their composition may be facilitated due to tribochemical reactions and/or chemical attack (corrosion). In particular, the chromium-nickel oxide-layer for stainless steel particles differed greatly from the oxide-layer of bulk implants. It needs further analysis, whether nickel in this oxide layer is bound as NiO_x or occurs as free Ni^0. If the latter is true, it will go quickly into solution as Ni^{2+}. In addition, as can be seen from Table 5, nickel-chlorides (NiC_{l2}, NiC_{l2} x 6 H_2O) may be present.

Conclusion

In summary, a methodology to fabricate titanium and stainless steel wear particles for experimental use has been presented. The morphology is comparable to characterized wear particles from in vivo and in vitro investigations. The chemical composition of wear particles may differ compared to the bulk material.

Future in vivo and in vitro studies utilizing these and retrieved wear particles are necessary to better understand the characteristic cell reactions of the artificially generated wear debris.

References

[1]   Malchau, H., Herberts, P., Soderman, P., and Oden, A., "Prognose der totalen Hueftarthroplastik. Aktualisierung und Validierung der Daten des Schwedischen Nationalen Hueftarthroplastik-Registers 1979-1998," 67th Annual Meeting American Academic of Orthopedic Surgeons, Orlando, FL, 2000, pp. 1–16.

[2]   Willert, H. G., Buchhorn, G. H., and Semlitsch, M., "Particle Disease Due to Wear of Metal Alloys. Findings from Retrieval Studies," *Biological, Material, and Mechanical Considerations of Joint Replacement*, B. F. Morrey, Ed., Raven Press, New York, 1993, pp. 129–146.

[3]   Urban, R. M., Jacobs, J. J., Gilbert, J. L., Rice, S. B., Jasty, M., Bragdon, C. R., et al., "Characterization of Solid Products of Corrosion Generized by Modular-Head Femoral Stems of Different Designs and Materials," *Modularity of Orthopeadic Implants, ASTM STP 1301*, D. E. Marlowe, J. E. Parr, and M. B. Mayor, Eds., ASTM International, West Conshohocken, PA, 1997, pp. 33–44.

[4]   Jacobs, J. J., Roebuck, K. A., Archibeck, M., Hallab, N. J., and Glant, T. T., "Osteolysis: Basic Science," *Clin Orthop*, Vol. 393, 2001, pp. 71–77.

[5]   Gilbert J. L. and Jacobs J. J., "The Mechanical and Electrochemical Processes Associated with Taper Fretting Crevice Corrosion: A Review," *Modularity of Orthopedic Implants, ASTM STP 1301*, D. E. Marlowe, J. E. Parr, and M. B. Mayor, Eds., ASTM International, West Conshohocken, PA, 1997, pp. 45–59.

[6]   Jones, D. M., Marsh, J. L., Nepola, J. V., Jacobs, J. J., Skipor, A. K., Urban, R. M., et al., "Focal Osteolysis at The Junctions of A Modular Stainless-Steel Femoral Intramedullary Nail," *J Bone Joint Surg Am*, Vol. 83, 2001, pp. 537–548.

[7]   Sommer, B., Ganz, R., Leunig, M., Guenther H. L., Felix, R., and Hofstetter, W. "Orthopaedic Materials Affect Osteoclast Formation," *Eur Cell Mater*, Vol. 1, Suppl. 2, 2001, p. 47.

[8]   Huo, M., Salvati, E., Lieberman, J., Betts, F., Bansal, M., and Oden, A., "Metallic Debris in Femoral Endosteolysis in Failed Cemented Total Hip Arthroplastics," *Clin Orthop*, Vol. 276, 1992, pp. 157–168.

[9] Shanbhag, A. S., Jacobs, J. J., Black, J., Galante J. O., and Glant, T. T. "Macrophage/Particle Interactions: Effect of Size, Composition and Surface Area," *J Biomed Mater Res*, Vol. 28, 1994, p. 81–90.

[10] Sprecher, C. M., Schneider, E., and Wimmer, M. A., "Generalized Size and Shape Description of UHMWPE Wear Debris - A Comparison of Cross-Linked, Enhanced Fused, and Standard Polyethylene Particle," *J ASTM Int*, Vol. 1, 2004, pp. 1–11.

[11] Cook, S. D., Renz, E. A., Barrack, R. L., Thomas, K. A., Harding, A. F., Haddad, R. J., et al., "Clinical and Metallurgical Analysis of Retrieved Internal Fixation Devices," *Clin Orthop*, Vol. 194, 1985, pp. 236–247.

[12] Lee, J. M., Salvati, E. A., Betts, F., DiCarlo, E. F., Doty, S. B., and Bullough, P. G., "Osteolysis: Basic Science," *J Bone Joint Surg Br*, Vol. 74, 1992, pp. 380–384.

[13] Urban, R. M., Jacobs, J. J., Tomlinson, M. J., Gavrilovic, J., Black, J., and Peoc'h, M., "Dissemination of Wear Particles to the Liver, Spleen, and Abdominal Lymph Nodes of Patients with Hip or Knee Replacement," *J Bone Joint Surg Am*, Vol. 82, 2000, pp. 457–476.

[14] Green, T. R., Fisher, J., Stone, M., Wroblewski, B. M., and Ingham, E. "Polyethylene Particles of a 'Critical Size' Are Necessary for the Induction of Cytokines by Macrophages in Vitro," *Biomaterials*, Vol. 19, 1998, pp. 2297–2302.

[15] Eschbach, L., Marti, A., and Gasser, B., "Fretting Corrosion Testing of Internal Fixation Plates and Screws," *Materials for Medical Engineering*, H. Stallforth and P. Ravell, Eds., Wiley-VCH, 1999, pp. 193–198.

Journal of ASTM International, January 2006, Vol. 3, No. 1
Paper ID JAI12817
Available online at www.astm.org

Nadim James Hallab,[1] *Shelly Anderson,*[1] *Marco Caicedo,*[1] *and Joshua J. Jacobs*[1]

Zirconium and Niobium Affect Human Osteoblasts, Fibroblasts, and Lymphocytes in a Similar Manner to More Traditional Implant Alloy Metals

ABSTRACT: Implant debris remains the major factor limiting the longevity of total joint replacements. Whether soluble implant debris of Zr and Nb containing implant alloys constitute a greater risk than other implant metals remains unknown. We evaluated the relative effects of soluble forms of $Zr+4$ and $Nb+5$ (0.001-10.0 mM) relative to $Cr+3$, $Mo+5$, $Al+3$, $Co+2$, $Ni+2$, $Fe+3$, $Cu+2$, $Mn+2$, $Mg+2$, $Na+2$, and $V+3$ chloride solutions on human peri-implant cells (i.e., osteoblast-like MG-63 cells, fibroblasts, and lymphocytes). Metals were ranked using a 50 % decrease in proliferation and viability to determine toxic concentrations. Lymphocytes, fibroblasts, and osteoblasts were, generally, similarly affected by metals where the most toxic metals, Co, Ni, Nb, and V required <1.0 mM to induce toxicity. Less toxic metals Al, Cr, Fe, Mo, and Zr generally required >1.0 mM challenge to produce toxicity. Overall, Co and V were the most toxic metals tested, thus Zr and Nb containing implant alloys would not likely be more toxic than traditional implant alloys. Below concentrations of 0.1 mM, neither Zr nor Nb reduced osteoblast, lymphocyte, or fibroblast proliferation. Zr was generally an order of magnitude less toxic than Nb to lymphocytes, fibroblasts, and osteoblasts. Our results indicated that soluble Zr and Nb resulting from implant degradation likely act in a metal- and concentration-specific manner capable of producing adverse local and remote tissue responses to the same degree as metals from traditional implant alloys, e.g., Ti-6Al-4V (ASTM F 138) and Co-Cr-Mo alloys (ASTM F 75).

KEYWORDS: zirconium, niobium, osteoblasts, lymphocytes, fibroblasts, metal ions, toxicity

Introduction

It remains unknown whether soluble metal debris from Zr and Nb containing implant alloys have a greater potential than traditional implant alloys to induce untoward effects in the peri-implant space. While particle induced inflammation remains the dominant mechanism compromising the integrity of the bone-implant interface through peri-implant bone loss (i.e., osteolysis) [1], there have been numerous reports that demonstrate an association between tissue necrosis and implant failure [2–4]. This is pertinent to clinical situations where large quantities of metal debris are generated [3,4]. Metallic debris generated through wear and corrosion can exist as micrometer to nanometer size particles and as soluble forms of metal (specifically or non-specifically bound to proteins) [5–12]. Are metals such as Zr and Nb, in implant alloys, preferentially toxic when compared to more traditional implant metals [13,14–19]? To date, the relative in vivo effects of soluble metals released from metallic implants on cells in the periprosthetic milieu remain largely unknown.

To investigate the potential of Zr and Nb soluble metals on peri-implant cells, we hypothesized that elevated concentrations of Zr and Nb will not adversely affect the function of

Manuscript received 13 September 2004; accepted for publication 12 July 2005; published January 2006. Presented at ASTM Symposium on Titanium, Niobium, Zirconium, and Tantalum for Medical and Surgical Applications on 9-10 November 2004 in Washington, DC.
[1] Department of Orthopedic Surgery, Rush University Medical Center, Chicago, IL 60612.

248

peri-implant cells to a greater degree than more traditional implant metals. We tested this hypothesis by treating representatives of human peri-implant cell types (i.e., osteoblast-like MG-63 cells, and primary human fibroblasts with 0.05–10 mM of Al^{3+}, Co^{2+}, Cr^{3+}, Fe^{3+}, Mo^{5+}, Nb, Ni^{2+}, V^{3+}, Zr^{4+}, Nb^{5+}, and Na^{2+} (control) chloride solutions. The degree to which soluble metals affected cell function was assessed using viability assays and proliferation assays. In this fashion, we determined which cell types were more affected by Zr and Nb implant metals and which overall metals were most toxic.

Materials and Methods

Cells

There were three types of cells used in this investigation, chosen to represent bone, soft tissue, and immune cells found in the peri-implant region: osteoblasts (human osteoblast-like cell line MG-63 (ATCC, Manassas, VA), fibroblasts (primary human fibroblasts (19 yr old female, ATCC, Rockville, MD)), and lymphocytes (Jurkat, T-helper lymphocytes, ATCC, Rockville, MD). All cells were used between 5–12 passages. The cells were cultured in Dulbecco's modified Eagle medium, DMEM (Sigma, St Louis, MO) at 37°C and 0.5 % CO_2, containing 10 % fetal bovine serum (FBS; Hyclone Laboratories, Inc., Logan, UT). MG-63 cells were cultured at eight concentrations of Al^{3+}, Co^{2+}, Cr^{3+}, Fe^{3+}, Mo^{5+}, Nb, Ni^{2+}, V^{3+}, Zr^{4+}, Nb^{5+}, and Na^{2+} (control) chloride solutions (Sigma, St Louis, MO) at 0.0, 0.01, 0.05, 0.1, 0.5, 1.0, 5, and 10.0 mM (Note: For MG-63 osteosarcoma cells, only Zr and Nb were newly tested for this investigation. The results of other metal chlorides for MG-63s were reported earlier [35]). Despite its prominence as a metal comprising implant alloys, titanium was not tested due its insoluble nature at physiologic pH (Ti precipitates as TiO_2 at a pH > 1). Corrosion products of metallic implants do not typically form single metal chlorides; thus the use of these metal challenge agents represent a baseline for further investigation (see Discussion).

Proliferation Assays

Proliferation of cells was measured using [^{3}H]-thymidine (Amersham International, Arlington Heights, IL) incorporation into DNA. Proliferation assays were performed using lymphocytes, fibroblasts, and osteoblasts cultured in 96-well cell-culture plates (Sigma), at a density of approximately 0.15×10^6 cells/well lymphocytes and $0.15–0.5 \times 10^5$ cells/well fibroblasts and osteoblasts for 48 h in 150 µL of DMEM/well, 10 % FBS at 37°C and 0.5 % CO_2, with or without metal treatments. Each metal concentration was tested in quadruplicate (4 wells/metal concentration) yielding a total of 384 samples for each cell type. [^{3}H]-thymidine (1 µCi /well) was added during the last 12 h of the 48-h culture period for osteoblasts and fibroblasts. However, for lymphocytes [^{3}H]-thymidine was added during the last 12 h of a 6 day incubation, the exposure time necessary for measuring cell mediated proliferation responses. This different time period was required to measure any Delayed Type Hypersensitivity (DTH) responses associated with metal exposure to Zr and Nb [20]. Incorporated radioactivity was measured using liquid scintillation Beta plate analysis (Wollac Gatesburg, MD). The amount of [^{3}H]-thymidine incorporation for each metal treatment was normalized to that of media alone (controls) producing a ratio, referred to here as the proliferation index or stimulation index, SI. The stimulations index was calculated using measured radiation counts per minute (cpm) as follows: Stimulation Index, SI = (mean cpm with treatment) / (mean cpm media alone). Proliferation assays using radioactive [^{3}H]-thymidine incorporation were used as a specific measure of proliferation, because the measured incorporation of thymidine into cellular DNA upon mitosis occurs only during the period of thymidine exposure (typically the last 8–24 h of challenge) and may not necessarily detect viable yet non-dividing cells. Viability assays were conducted to measure toxicity more directly. Proliferation assay results were categorized using a >50 % decrease in proliferation, IC_{50} (half inhibitory concentration) index, to compare metals. All proliferation testing was conducted in quadruplicate but was limited to one time point.

Viability Assays

Cell viability was determined by: 1) trypan blue staining using a hemacytometer and 2) flow cytometry (FACScan, Becton Dickinson Co., San diego, CA) using propidium iodide (Calbiochem, San Diego, CA) staining. Osteoblast and fibroblast viability was determined using trypan blue exclusion and hemacytometry of all four quadrants. Lymphocyte non-adherence facilitated viability testing using flow cytometry assays. Flow cytometry viability assays assessed plasma membrane integrity using propidium iodide, a fluorescent vital dye that stains nuclear DNA. Propidium iodide does not cross the plasma membrane of viable cells. The propidium iodide stock solution was composed of 50 μg propidium iodide /ml in PBS (at pH 7.4) for flow cytometric assays, where 10 μl of this stock solution was used per 1×10^5 cells after 48 h incubation with each of the metal solutions at each of the eight concentrations. General viability/toxicity was evaluated using a LC50 index (half lethal concentration, i.e., the concentration at which 50 % of the cells were viable) to compare metals. All viability tests were conducted in triplicate.

Statistical Analysis

Normally distributed data (proliferation and viability data) were subjected to statistical analysis using Student's t-tests. Student's t-tests for independent samples with unequal or equal variances were used to test equality of the mean values ($p<0.05$). Comparisons between groups are limited to individual comparison of reactivity for each metal concentration.

Results

Proliferation Assays

Proliferation assays of metal challenged osteoblasts, fibroblasts, and lymphocytes demonstrated differences of up to 3 orders of magnitude among the metals tested (Figs. 1–3). While Zr was less inhibitory to all three cell types than Nb, both Zr and Nb were less inhibitory to all three cell types than either Co or V. Lymphocyte was the cell type most tolerant of Zr and Nb challenge, where approximately 10 mM were required to induce an IC_{50}, compared to the approximately 1 mM dose which inhibited osteoblast and fibroblast proliferation by >50 %.

Osteoblasts

Nb and Zr were similarly inhibitory to MG-63 osteoblasts, both requiring 5 mM to induce a 50 % decrease in proliferation (Fig. 1). Additionally, both Zr and Nb demonstrated no toxic effect on osteoblast proliferation at up to 1 mM. The three most osteo-inhibitory metals, Co, Ni, and V, reduced osteoblast proliferation by 50 % at <1 mM, at a dose an order of magnitude less than that of Zr or Nb. The most inhibitory metal was V where a dose of 0.1 mM induced a 50 % decrease in osteoblast proliferation(IC_{50}), an order of magnitude less than all other metals. All metals induced 50 % reductions in proliferation by 10 mM challenge concentrations.

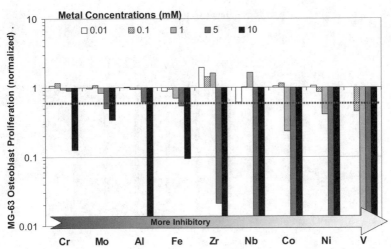

FIG. 1—*Normalized MG-63 osteoblast proliferation responses to metal challenge after 48 h of exposure are shown from more to less inhibitory (left to right). The most toxic (inhibitory) metal was V, where a dose of 0.1 mM induced a 50 % decrease in osteoblast proliferation, as indicated by dashed line. The other metals required an order of magnitude greater concentration to induce a 50 % decrease, i.e., Co and Ni at approximately 1.0 mM. The least inhibitory metals (Fe, Al, Mo, and Cr) required 10 mM concentration to induce a 50 % decrease in proliferation.*

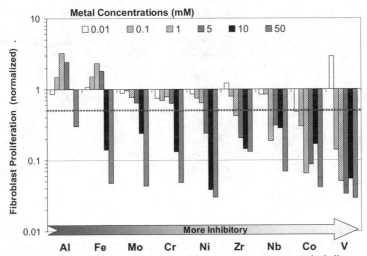

FIG. 2—*Normalized primary human fibroblast proliferation responses to metal challenge after 48 h of exposure. Significantly increased proliferation responses were induced by Al, V, and Fe at specific concentrations. Aluminum induced an increased fibroblast proliferation response at 0.1–5.0 mM and required a concentration of approximately 50 mM to decrease proliferation to <50 %. The most inhibitory metals were Co and V (e.g., Stimulation Index <0.5 at 0.01 mM Co and 0.1 mM V), which induced an IC_{50} at concentrations an order of magnitude below that of other metal challenge agents. The metals found to be less inhibitory to fibroblast proliferation were Mo, Cr, Fe, and Al, which all required > 5mM to reduce fibroblast proliferation by greater than 50 %.*

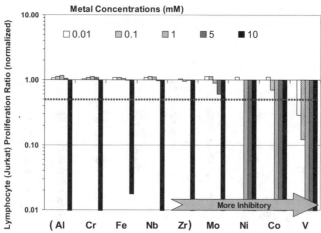

FIG. 3—*Jurkat lymphocytes (lymphoma derived human T-helper cells) demonstrated only inhibited proliferation to metal challenge. The metal that decreased proliferation at the least concentration was V.*

Fibroblasts

Nb and Zr inhibited fibroblast proliferation by 50 % at a concentration of <1 mM, which was less inhibitory than both Co and V (Fig. 2). The IC_{50} of both Co and V (i.e., at 0.01 mM Co and 0.1 V) were an order of magnitude below that of other metal challenge agents, making them the most inhibitory soluble metal challenge agents on fibroblasts. The metals found to be least inhibitory to fibroblast proliferation were Al, Cr, Fe, Mo, and Ni, which all required ≥5 mM to reduce fibroblast proliferation by greater than 50 %. Zr and Nb were more inhibitory than these metals (IC_{50} at 1 mM) and thus were categorized as moderately inhibitory to fibroblasts.

Lymphocytes

Jurkat cells (human T-helper cell line) generally demonstrated greater tolerance to metal challenge than fibroblasts or osteoblasts, where the least inhibitory metals, Al, Cr, Fe, Nb, and Zr, required >5 mM to decrease lymphocyte responses by 50 % (Fig. 3). Here Mo, Ni, Co, and V were more inhibitory to lymphocytes than Zr or Nb by approximately an order of magnitude. Al, Cr, Fe, Nb, and Zr metals did not significantly reduce lymphocyte proliferation at <5 mM. Again, as was the case with osteoblasts and fibroblasts, Co and V were the most lympho-inhibitory metals (with <1 mM metal challenge reducing cell proliferation by >50 %), with V an order of magnitude more inhibitory than Co.

Viability Assays

The viability of osteoblasts, fibroblasts, and lymphocytes in metal solutions was tested over a range of metal concentrations (0.01–10 mM) using flow cytometry or trypan blue-hemacytometry assessment of cell membrane integrity. Metal-induced effects were categorized as toxic at a 50 % viability (LC50), and all occurred above 0.1 mM. The metals tested induced toxicity within three orders of magnitude in concentration, where each metal was categorized as mild, moderate, or highly toxic. Thus, metals were categorized as "highly toxic" if they induced toxicity within the approximate range of 0.01–0.5mM, "moderately toxic" if they induced toxicity within the approximate range of 0.5–5.0 mM concentrations, and "mildly toxic" if they induced toxicity at concentrations above 5 mM (Figs. 4–6).

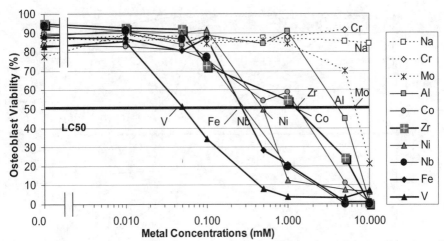

FIG. 4—*Soluble metal challenge-induced toxicity on MG-63 osteoblasts is shown. Below concentrations of 0.01 mM, none of the metals reduced osteoblast viability. Based on a 50 % reduction in viability (LC50) the most toxic metals (black symbols), were Fe, Nb, and V at <0.5 mM; the moderately toxic metals (gray symbols) were Ni, Zr, Co, and Al at 0.5–5.0 mM; and the mildly toxic metals (open symbols) were Cr and Mo at >5.0 mM.*

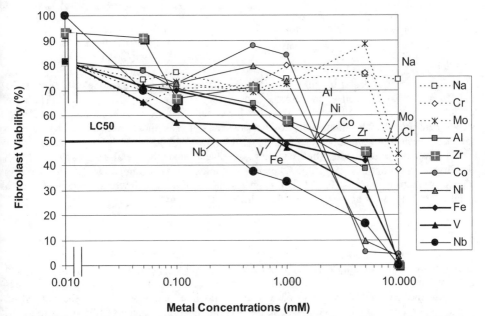

FIG. 5—*Metal challenged fibroblast toxicity is shown (symbols are similar to those used for osteoblasts to facilitate comparison). The most fibroblast-toxic metals (i.e. Nb, V, and Fe) reduced cell viability at concentrations <1.0 mM. Moderately toxic metals were those that reduced viability to <50 % between approximately 1–5.0 mM (i.e., Al, Co, Ni, and Zr). The remaining metals (Cr and Mo) were characterized as mildly toxic to fibroblasts since they retained greater than 50 % viability at concentrations >5.0 mM.*

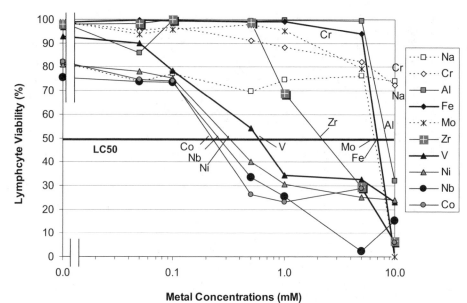

Metal Concentrations (mM)

FIG. 6—*Viability testing on Jurkat lymphocytes was conducted using metal challenge agents at 6 concentrations. Metals that were moderately toxic to osteoblasts and fibroblasts (Co, Nb, and Ni) were the most toxic metals to lymphocytes, where the LC50 of Co and Ni were 0.2 and 0.3 mM, respectively.*

Osteoblasts

Below concentrations of 0.1 mM, none of the metals reduced osteoblast viability (Fig. 4). Zr and Nb were characterized as moderately and highly osteoblast-toxic, respectively. The most "toxic" metals were Fe, Nb, and V, which reduced cell viability at concentrations <0.5 mM. Metals determined to be moderately toxic to osteoblasts (i.e., those that reduced viability to <50 % between approximately 0.5–5.0 mM) were Ni, Co, Zr, and Al. The remaining metals (Cr and Mo) were characterized as mildly toxic to osteoblasts since they retained greater than 50 % viability at concentrations >5.0 mM. As expected, Na (as NaCl) was nontoxic (>50 % viability) at concentrations as high as 10 mM. This suggests that chloride, which was a component of each metal salt solution, was not responsible for inducing cell toxicity.

Fibroblasts

Metal challenged fibroblasts (48-h culture) demonstrated concentration-dependent toxicity responses similar to osteoblasts (Fig. 5). Because the same passage of human fibroblast cells and the same timeframe was used for testing (excluding Zr and Nb), the control value for all metals was the same. Zr and Nb were tested at a different time, and those fibroblasts retained an unchallenged (control) viability of >90 %. Nb was the most toxic metal to fibroblasts followed by V and Fe where all three were categorized as being the highly fibro-toxic metals reducing cell viability at concentrations <1.0 mM. Metals determined to be moderately toxic to fibroblasts were (at approximately 1–5.0 mM) Al, Co, Ni, and Zr. The remaining metals, Cr and Mo, were characterized as mildly toxic to fibroblasts since the cells retained greater than 50 % viability at concentrations >5.0 mM. Similar to osteoblasts, Na was nontoxic to fibroblasts (>50 % viability) at concentrations as high as 10 mM. Except for Nb, fibroblasts generally demonstrated greater generalized tolerance to metal challenge than osteoblasts in that the metal challenge agents Nb, V, Fe, Al, Ni, Zr, and Co produced a 50 % decrease in fibroblast viability between the concentrations of 1–5 mM, an order of magnitude greater than tolerated by osteoblasts.

Lymphocytes

Viability testing on lymphocytes was conducted using Jurkat (lymphoma derived human T-helper) cells (Fig. 6). The variability of "controls" viability (untreated lymphocytes with 80 to 100 % viability) is reflected in Fig. 6 and is inherent to using separate controls for each metal because each was conducted at a different time or cell passage. Zr and Nb were similarly toxic to lymphocytes as they were to osteoblasts and fibroblasts. The viability responses of lymphocytes to metal challenge were similar to those demonstrated in the proliferation assays, where the most lymphotoxic metals were Co, Nb, Ni, and V, which induced >50 % decrease in viability at concentrations <1.0 mM. The majority of metals tested (Al, Cr, Fe, Mo, and Zr) were only mildly lymphotoxic where a decrease in viability to <50 % was not induced at concentrations below 5.0 mM. The most lymphotoxic metal was Co, where a concentration of approximately 0.2 mM resulted in a >50 % decrease in lymphocyte viability. Nb was only slightly less lymphotoxic than Zr. Iron, a more toxic metal to fibroblasts and osteoblasts, was relatively non-toxic to lymphocytes over a wide range of challenge concentrations (0.05–1 mM).

Discussion

The concentrations at which Zr and Nb metals negatively affected peri-implant cells (i.e., lymphocytes, fibroblasts, and osteoblasts) in this investigation support our hypothesis that Zr and Nb alloys do not preferentially (to traditional implant metals) induce cell toxicity. This limited yet novel finding is important given the increasing amount of biologic testing that shows potential to tailor implant material selection to best suit an individual's reactivity profile, where additional implant alloys increase the armamentarium of orthopedic surgeons. Zr and Nb seem to be less toxic than Co and V, which were previously found to be the most toxic implant metals to osteoblasts [35]. Co and V generally reduced the proliferation and viability of peri-implant by >50 % cells at concentrations <1.0 mM. Generally, the three representative cell lines were similarly affected by Zr and Nb, where Nb was more toxic than Zr. This indicates preferential cells responses, e.g., immunosuppression (via toxicity or decreased proliferation), is not likely without concomitant generalized tissue necrosis (concomitant fibroblast and osteoblast metal toxicity).

Previous studies agree (rather predictably) that metals can negatively impact fibroblasts, endothelial cells, and non-human osteoblast-like cells [21–34]. Some have shown potential for metal-induced inflammatory responses [35–37] and suppressed bone formation [21]. Zr and Nb toxicity has not been previously studied (or comparatively or otherwise) using a wide range of concentrations and cell types. We previously demonstrated the utility of this soluble metal proliferation-assay-centric model to support cell-mechanism dysfunction, i.e., MG-63 osteoblasts demonstrated depressed collagen production was concomitant with decreased proliferation to Cr, Co, and Ni soluble challenge [35].

Different metals induced different peri-implant cell toxicity. Based on metal dose required to induce a >50 % reduction in proliferation and viability, Co, Ni, Nb, and V tended to be more toxic (IC_{50} and LC50 <1 mM) to lymphocytes, fibroblasts, and osteoblasts, whereas Cr, Mo, Al, and Zr were generally less toxic. Implant alloys containing these more toxic metals (Co, Ni, and V) currently include the three most prevalent implant alloys, Ti-alloy (4 %V), Co-alloy (61–66%Co and <2 %Ni), and stainless steel (10–15 % Ni). Past investigations suggest that Co-alloy soluble degradation products are more likely to be increased in serum compared to Ti-alloy components [16,38–40]. It remains unknown how Zr and Nb will be metabolized (retained/excreted) when released from orthopaedic implants, and thus no in vivo comparison can be made to the levels determined to be toxic in this investigation.

Limitations inherent to this line of investigation include: (1) the appropriateness of the challenge conditions and (2) an incomplete understanding of what specific mechanism(s) mediate metal induced toxic effects on each cell type. The form(s) of soluble metal-implant debris in vivo that may contribute to adverse effects remains incompletely characterized. The presumption that soluble implant debris can be modeled using metal salt challenge agents is likely an oversimplification, given the complex dissolution chemistry of implant alloys in a complex electrolyte, such as serum, where soluble metals are bound specifically and non-specifically by serum proteins [41]. However, the error involved in using metal chlorides as proxies for corrosion products may be over-conservative when asserting that certain levels of metal exposure are or can be toxic. This study used the more stable and less toxic valences of each metal, e.g., tri-valent Cr rather than hexa-valent Cr to ascertain if the more benign forms of the common can affect cells. These less toxic metal compounds were all chlorides. Other more toxic forms and valences of metals (e.g., Cr^{6+}) are classified under the Official Journal of the European Communities, Annex Commission Directive 98/98/ECC. The use of human cell lines and in vitro conditions as an approximation of primary in vivo human cells is another limitation. However, the use of relatively resilient lymphoma and osteosarcoma cell lines that approximate in vivo cells are particularly useful in conservatively approximating upper limits of lymphocyte and osteoblast tolerance to soluble metals. The delicate nature of primary cells (osteoblasts more than lymphocytes) in tissue culture would likely inadequately represent compensatory environments in vivo. The approximations and limitations of this testing indicate the greatest utility of this testing may lie more in the general ranking of metals and comparison of cell resilience than the ability to determine concentrations of toxic metal exposure quantitatively. Further study is continuing using a multi-assay approach in an attempt to more thoroughly support a ranking of soluble metal effects on peri-implant cell proxies such as lymphocytes, fibroblasts, and osteoblasts. Important to this line of investigation is the inclusion of mixed metal challenge agents where cells are exposed to clinically relevant mixtures of soluble metal. However, subtle distinctions between nano-particulate and soluble debris and what exactly are the relative amounts of soluble metal debris in vivo remain incompletely characterized. The subtle differential susceptibility of MG-63 osteoblasts to metals such as Ni when compared to primary fibroblasts or lymphocytes would likely be further exaggerated when using primary osteoblasts, were it technically practical to obtain the large numbers required for the current investigation. However, cell type differences were not evident when examining the effects of Zr and Nb. Whether the complex environment of the peri-implant milieu ultimately mitigates or makes cells more differentially susceptible to the effects of Zr and Nb soluble metal exposure in vivo is unknown.

Conclusions

The response of proxy peri-implant cells to Zr and Nb metal challenge were similar to other implant alloy metals. Co and V were generally the most toxic metals to lymphocytes, fibroblasts, and osteoblasts. These three cell types were generally similarly affected by metal challenge, where the most toxic metals, Co, Ni, Nb, and V required <1.0 mM to induce toxicity. Less toxic metals Al, Cr, Fe, Mo, and Zr generally required >1.0 mM challenge to produce toxicity. Implant alloys containing the more toxic metals (Co, Nb, Ni, and V) currently include the three most prevalent implant alloys, Co-alloy (61–66 %Co and <2 %Ni), Ti-alloy (4 %V) and stainless steel (10–15 % Ni); thus toxicity concerns regarding Zr and Nb in newer alloys (e.g., Oxinium™, Zr-2.5 %Nb) seem less warranted than similar concerns of more traditional alloys. Currently, no clinical data are available to compare in vitro Zr and Nb levels demonstrated to produce toxicity with levels that occur as a result of alloy degradation in vivo. Some subtle metal-induced differences between cell types were observed. Nb induced the greatest decrease in fibroblast viability, but Co and V were more inhibiting to fibroblast proliferation. Fibroblast proliferation was less inhibited when challenged with Al, Fe, Cr, and Mo (at 5 mM) than were osteoblasts and lymphocytes. The concentrations of metal found toxic in this study may serve as conservative guidelines. These findings support the hypothesis that adverse local and remote tissue responses to Zr and Nb alloy implant metals are not likely to induce necrotic reactivity (toxic responses) differentially. Other mechanisms of debris induced implant failure, e.g., particle induced aseptic osteolysis, are not necessarily impacted by these results where implant alloy particles, not soluble debris, are purportedly responsible for bone resorbing inflammation. However, soluble forms of specific metal degradation products provide a necessary foundation for comparative evaluation of the effects of both soluble and particulate implant debris of newer implant alloy metals.

Acknowledgments

We would like to thank the NIH/NIAMS, the Crown Family Chair of Orthopedics, and the Rush Arthritis and Orthopaedics Institute for their support.

References

[1] Jacobs J, Goodman S, Sumner DR, Hallab N: Biologic response to orthopedic implants. pp. 402-426. Orthopedic Basic Science. American Academy of Orthopedic Surgeons; Chicago, 1999.

[2] Doorn PF, Mirra JM, Campbell PA, Amstutz HC: Tissue reaction to metal on metal total hip prostheses. Clin OrthopS187-S205, 1996.

[3] Campbell PA, McKellop H, Alim R, Mirra J, Nutt, S, Amstutz HC: Metal-on-metal hips replacements: Wear performance and cellular response to wear particles. pp. 193-200. In Disegi R, Kennedy R.L., Pilliar RM (eds): Cobalt-Base alloys for Biomedical Applications, ASTM STP 1365. ASTM International, West Conshohocken, PA, 1999.

[4] Svensson O, Mathiesen EB, Reinholt FP, Blomgren G: Formation of a fulminant soft-tissue pseudotumor after uncemented hip arthroplasty. A case report. J Bone Joint Surg Am 70:1238-1242, 1988.

[5] Hallab NJ, Jacobs JJ, Skipor A, Black J, Mikecz K, Galante JO: Systemic metal-protein binding associated with total joint replacement arthroplasty. J Biomed Mater Res 49:353-361, 2000.

[6] Wang JC, Yu WD, Sandhu HS, Betts F, Bhuta S, Delamarter RB: Metal debris from titanium spinal implants. Spine 24:899-903, 1999.

[7] Goodman SB, Lind M, Song Y, Smith RL: In vitro, in vivo, and tissue retrieval studies on particulate debris. Clin Orthop 352:25-34, 1998.

[8] Huo MH, Salvati EA, Lieberman JR, Betts F, Bansal M: Metallic debris in femoral endosteolysis in failed cemented total hip arthroplasties. Clin Orthop 276:157-168, 1992.

[9] Shanbhag AS, Jacobs JJ, Glant TT, Gilbert JL, Black J, Galante JO: Composition and morphology of wear debris in failed uncemented total hip replacement. J Bone Joint Surg Br 76:60-67, 1994.

[10] Jacobs JJ, Urban RM, Schajowicz F, Gavrilovic J, Galante JO: Particulate-associated endosteal osteolysis in titanium-base alloy cementless total hip replacement. Particulate Debris from Medical Implants. ASTM International, West Conshohocken, PA, 1992.

[11] Urban RM, Jacobs JJ, Gilbert JL, Galante JO: Migration of corrosion products from modular hip prostheses. Particle microanalysis and histopathological findings. J Bone Joint Surg [Am] 76:1345-1359, 1994.

[12] Urban RM, Jacobs JJ, Tomlinson MJ, Black J, Turner TM, Sauer PA, Galante JO: Particles of metal alloys and their corrosion products in the liver, spleen and para-aortic lymph nodes of patients with total hip replacement prosthesis. Orthop Trans 19:1107-1108, 1996.

[13] Dorr LD, Bloebaum R, Emmanual J, Meldrum R: Histologic, biochemical and ion analysis of tissue and fluids retrieved during total hip arthroplasty. Clinical Orthopedics and Related Research 261:82-95, 1990.

[14] Jacobs JJ, Skipor AK, Urban RM, Black J, Manion LM, Starr A, Talbert LF, Galante JO: Systemic distribution of metal degradation products from titanium alloy total hip replacements: an autopsy study. Trans Orthop Res Soc New Orleans:838, 1994.

[15] Michel R, Nolte M, Reich M, Loer F: Systemic effects of implanted prostheses made of cobalt-chromium alloys. Archives of Orthopaedic and Trauma Surgery 110:61-74, 1991.

[16] Stulberg BN, Merritt K, Bauer T: Metallic wear debris in metal-backed patellar failure. Journal of Biomed Mat Res Applied Biomaterials 5:9-16, 1994.

[17] Willert HG, Broback LG, Buchhorn GH, Jensen PH, Koster G, Lang I, Ochsner P, Schenk R: Crevice corrosion of cemented titanium alloy stems in total hip replacements. Clin Orthop 333:51-75, 1996.

[18] Jacobs J, Urban RM, Gilbert JL, Skipor A, Black J, Jasty MJ, Galante JO: Local and distant products from modularity. Clin Orthop 319:94-105, 1995.

[19] Urban RM, Jacobs JJ, Tomlinson MJ, Gavrilovic J, Black J, Peoc'h M: Dissemination of wear particles to the liver, spleen, and abdominal lymph nodes of patients with hip or knee replacement. J Bone Joint Surg [Am] 82:457-476, 2000.

[20] Cederbrant K, Hultman P, Marcusson JA, Tibbling L: In vitro lymphocyte proliferation as compared to patch test using gold, palladium and nickel. Int Arch Allergy Immunol 112:212-217, 1997.

[21] Sun ZL, Wataha JC, Hanks CT: Effects of metal ions on osteoblast-like cell metabolism and differentiation. J Biomed Mater Res 34:29-37, 1997.

[22] Wataha JC, Hanks CT, Sun Z: Effect of cell line on in vitro metal ion cytotoxicity. Dent Mater 10:156-161, 1994.

[23] Wataha JC, Hanks CT, Craig RG: In vitro effects of metal ions on cellular metabolism and the correlation between these effects and the uptake of the ions. J Biomed Mater Res 28:427-433, 1994.

[24] Wataha JC, Hanks CT, Craig RG: The effect of cell monolayer density on the cytotoxicity of metal ions which are released from dental alloys. Dent Mater 9:172-176, 1993.

[25] Nichols KG, Puleo DA: Effect of metal ions on the formation and function of osteoclastic cells in vitro. J Biomed Mater Res 35:265-271, 1997.

[26] Thompson GJ, Puleo DA: Ti-6Al-4V ion solution inhibition of osteogenic cell phenotype as a function of differentiation timecourse in vitro. Biomater 17:1949-1954, 1996.

[27] Thompson GJ, Puleo DA: Effects of sublethal metal ion concentrations on osteogenic cells derived from bone marrow stromal cells. J Appl Biomater 6:249-258, 1995.

[28] Puleo DA, Huh WW: Acute toxicity of metal ions in cultures of osteogenic cells derived from bone marrow stromal cells. J Appl Biomater 6:109-116, 1995.

[29] Morais S, Dias N, Sousa JP, Fernandes MH, Carvalho GS: In vitro osteoblastic differentiation of human bone marrow cells in the presence of metal ions [In Process Citation]. J Biomed Mater Res 44:176-190, 1999.

[30] Morais S, Sousa JP, Fernandes MH, Carvalho GS, de Bruijn JD, van Blitterswijk CA: Effects of AISI 316L corrosion products in in vitro bone formation. Biomater 19:999-1007, 1998.

[31] Morais S, Sousa JP, Fernandes MH, Carvalho GS: In vitro biomineralization by osteoblast-like cells. I. Retardation of tissue mineralization by metal salts. Biomater 19:13-21, 1998.

[32] Wagner M, Klein CL, van Kooten TG, Kirkpatrick CJ: Mechanisms of cell activation by heavy metal ions. J Biomed Mater Res 42:443-452, 1998.

[33] Edwards DL, Wataha JC, Hanks CT: Uptake and reversibility of uptake of nickel by human macrophages. J Oral Rehabil 25:2-7, 1998.

[34] Wang JY, Wicklund BH, Gustilo RB, Tsukayama DT: Prosthetic metals interfere with the functions of human osteoblast cells in vitro. Clin Orthop 339:216-226, 1997.

[35] Hallab NJ, Vermes C, Messina C, Roebuck KA, Glant TT, Jacobs JJ: Concentration- and composition-dependent effects of metal ions on human MG-63 osteoblasts. J Biomed Mater Res 60:420-433, 2002.

[36] Hallab NJ, Messina C, Stafford T, Jacobs JJ: Metal specific responses dominate in vitro hypersensitivity reactions in patients with total joint arthroplasty. Trans 49th Orthopaedic Research Society 28:286, 2003.

[37] Hallab NJ, Skipor A, Jacobs JJ: Interfacial kinetics of titanium- and cobalt-based implant alloys in human serum: metal release and biofilm formation. J Biomed Mater Res 65A:311-318, 2003.

[38] Jacobs JJ, Gilbert JL, Urban RM: Corrosion of metal orthopaedic implants. J Bone Joint Surg [Am] 80:268-282, 1998.

[39] Sunderman FW, Hopfer SM, Swift T, Rezuke WN, Ziebka L, Highman P, Edwards B, Folcik M, Gossling HR: Cobalt, chromium, and nickel concentrations in body fluids of patients with porous-coated knee or hip prostheses. J Orthop Res 7:307-315, 1989.

[40] Black J, Maitin EC, Gelman H, Morris DM: Serum concentrations of chromium, cobalt, and nickel after total hip replacement: a six month study. Biomater 4:160-164, 1983.

[41] Hallab NJ, Jacobs JJ, Skipor A, Black J, Mikecz K, Galante JO: Systemic metal-protein binding associated with total joint replacement arthroplasty. J Biomed Mater Res 49:353-361, 2000.

Author Index

Subject Index

V

Vanadium, 248

W

Wear debris, 239, 248

X

X-ray photoelectron spectroscopy, 225

Y

Young's modulus, titanium alloy, 135

Z

Zirconium alloy, 248
  bacterial adhesion, 225
Zirconium othopedic implants, 16